J. Pickering (John Pickering) Putnam

Lectures on the Principles of House Drainage

J. Pickering (John Pickering) Putnam

Lectures on the Principles of House Drainage

ISBN/EAN: 9783743686779

Printed in Europe, USA, Canada, Australia, Japan

Cover: Foto ©berggeist007 / pixelio.de

More available books at **www.hansebooks.com**

LECTURES

ON

THE PRINCIPLES OF HOUSE DRAINAGE,

DELIVERED BEFORE THE

SUFFOLK DISTRICT MEDICAL SOCIETY
(Section for Clinical Medicine, Pathology, and Hygiene),

AND THE

BOSTON SOCIETY OF ARCHITECTS,

AT THE

MASS. INSTITUTE OF TECHNOLOGY.

BY

J. PICKERING PUTNAM,
ARCHITECT.

*Part 1 reprinted from the Boston Medical and Surgical Journal.
Nos. 17, 18, and 19, 1885.*

From the Editor of the *Sanitary Record*, London.
By permitting me to reproduce these "admirable articles in the *Sanitary Record*, you will greatly oblige me and **serve the cause of** Sanitary Science in England."   THE EDITOR.

BOSTON:
TICKNOR AND COMPANY.
1886

PRESS OF
STANLEY AND USHER,
BOSTON.

THE PRINCIPLES OF HOUSE DRAINAGE.

PART I.

[*Reprinted from the Boston Medical and Surgical Journal.*]

As announced in the invitation cards to this lecture, a number of the appliances used for illustration are inventions of my own, and some of them are patented. These devices have been clearly marked on the drawings with their name, " Sanitas," and in referring to them they will always be so designated. The object of this is that every one may know when anything of my own is referred to, and be able to assure himself that the subject is treated without bias. My intention is to make no statement which is not founded on facts recognized by all or easily demonstrated, nor to follow any course of reasoning which is not perfectly clear and logical. In order, moreover, to make our meeting as satisfactory and fruitful as possible, it is hoped that every one will make a note of any point which may not be perfectly clear to him, or of any deductions with which he may not be fully in accord, and mention them in the discussion following the lecture, so that the reasoning leading to these deductions may be reëxamined or more clearly stated.

It is proper to add that the devices are the outgrowth of a careful practical study of plumbing made from the unprejudiced standpoint of the architect working for the interest of his client.

They **are** the *result*, **and not** the *cause*, of **the** investigations.

As plumbing **is now practised, the** architect or the sanitary engineer **is, from** the nature of his **work,** the **one upon whom we must** depend chiefly **for its improvement.** Evidently **the most important** part **of an** architect's work **is** that **which concerns the health and** comfort of his client.

The arrangement of the plumbing pipes **and fixtures** influences the **entire** plan **from** foundation **to roof.**

Some of the rooms, **such as the laundry and toilet rooms, are designed exclusively for the** plumb**ing,** and **all are more or less dependent upon its** arrangement. The walls and beams must **be slotted** and framed for its **reception,** and differently **for each** different kind of fixture **or** system **of** piping **and for** their lighting and **ventilating** apparatus. Hence, the architect must be **familiar** with all **the** details of the work, and **upon** him **lies the responsibility, not** only **for the** healthfulness, **convenience, and cost of** the particular work over which **he** has immediate charge, but also, in a great measure, for the general status of the **art of** plumbing throughout the country. The plumber, like **the** architect, **may see** defects in the methods of plumbing now in vogue, **but he has** comparatively little **interest in** promoting **reform** because **the** authority lies **with** the architect.* **The plumber has contracted to do a** certain **amount of work for a certain amount of** money **and it is** not easy **for him to alter the** contract. **If he is directed to put into a house** a few hundred dollars' worth **of** piping more than

* Some architectural firms now employ regularly a Sanitary Engineer, either directly or indirectly connected with their offices, to take charge, in coöperation with the architect, of the department of heating, ventilation, and plumbing. The custom has much to recommend in it.

is necessary or desirable, or set complicated or ill-devised fixtures, it is not his duty to protest. Competition has reduced his profits to so small a figure that the plumber cannot *afford* to be an active reformer.

The physician seldom interferes in the details of plumbing work, and the public are, as a rule, profoundly ignorant of them.

The architect, however, acting directly as the agent of the owner, is bound, when he discovers defects affecting the health as well as the pocket of his client, to use all the influence he possesses to remove them.

Every sanitarian recognizes the serious defects of our present plumbing methods and apparatus. Our common soil and waste pipes are neither gas nor water tight on account of their defective form and jointing, and they are expensive to lay.

Our traps either are incapable of retaining their water-seal against the adverse influences affecting them in common use, or they are bulky, unscientific, and expensive.

Our lavatories are slow emptying, inconvenient, and complicated. Most of the water-closets in use are full of defects and flushed on incorrect principles, and, in short, there is scarcely a single point in which our plumbing is not evidently susceptible of great improvement.

Let us examine these defects and study the principles which should guide us in effecting their cure : —

Simplicity. The tendency at present is toward undue complication. The plumbing work is becoming each year more elaborate and costly, more difficult to set correctly, and more difficult to comprehend and repair when correctly set, so that the public are becoming alarmed and confused. They

despair of being able to understand the intricate
system of piping and machinery for the supply and
waste of fixtures. The result is a general feeling
of insecurity and a tendency to forego the convenience of plumbing fixtures wherever their presence
is not an absolute necessity.

Our byword should be "simplicity." Rather
than reduce the number of our fixtures, let us reduce
the amount of machinery connected with them, provided we can do so without diminishing the security
they are intended to afford.

If we find that our process of simplification
actually increases the security of the work, all will
be gainers — the public as well as the plumber.
For what is best for the public by increasing their
confidence is also best for the plumbers, though they
appear often to lose sight of this fact.

Accessibility. Another leading principle is that
all plumbing work in a house should be everywhere,
without exception, accessible and as far as possible
visible.

Pipes should never run behind plaster when it is
possible to expose them on walls and ceilings. The
pipes both waste and supply as well as the bathtub
traps of a bathroom should be placed, if possible,
on the ceiling underneath the plaster of the bathroom or china-closet below — never between the floor-
joists. There is nothing in a neatly arranged line
of lead, brass, and iron piping that one needs to be
ashamed of. On the contrary when skilfully placed
and neatly jointed in a workmanlike manner, as
would be the case when the plumber knew they were
to be forever exposed to view, these bright metal
pipes become quite ornamental when mounted with
lead or brass clamps on strips of finished woodwork
varnished and symmetrically arranged in corners
or where good taste and judgment direct.

In the darker ages of architectural art, chimneys were despised and hidden from view. Now they become the most prominent features of a design, suggesting hospitable comfort and healthful ventilation within. So should it be with the piping. A knowledge on the part of the house-owner that all the pipes which provide him with the comforts of pure water and safely carry off the foul are in full view and in a sound condition will afford him much solid satisfaction.

Avoidance of Mechanical Obstructions. A third principle is to avoid all mechanical obstructions, such as valves, balls, gates, and all other impediments to the water-way, and in a system of water-carriage to do all trapping by means of a water-seal alone.

Mechanical devices form no reliable security against the passage of sewer gas. These valves and balls cannot be made to fit their seats with such accuracy as to exclude liquids and gases, or microscopic disease-germs, even when new. They soon become more or less fouled with dirt and corrosion and then their inefficiency becomes evident even to the eye. A sound water-seal, however, properly protected, is found to be entirely reliable in excluding noxious matters of all kinds.

Fig. 1 represents a trap having the undesirable mechanical seal in the form of a ball.

Moreover we are *obliged* to rely upon a simple water-seal whether we desire to or not, because our water-closet traps or their overflows are, and must be, constructed without mechanical obstructions. Evidently if the water-seal is inefficient we must either construct our water-closet

Fig. 1. The Jennings trap, with mechanical seal.

traps and their overflows on a different principle
or else give up the system of water-carriage altogether. It is useless to apply mechanical closures
to our smaller traps if we leave the large water-
closet traps without them.

It has been shown by Dr. Carmichael and others
that if a water-seal be properly maintained against
evaporation and siphonage, or destruction from any
cause, the amount of sewer gas that can pass
through in twenty-four hours, even under the worst
conditions, but with a ventilated soil-pipe, is infini-
tesimal and absolutely harmless, and that disease-
germs cannot pass at all through water at rest at
normal temperatures. Dr. Carmichael also experi-
mented with an unventilated soil-pipe, and found
here that the quantity of carbonic-acid gas, the
largest component of sewer gas, given off from the
traps in twenty-four hours was less than that
obtained " when a bottle of lemonade was opened,"
and less than that which is exhaled by a man in
five minutes.

As for the ammonia, sulphuretted hydrogen, and
other gases or vapors which accompany the carbonic
acid, their combined amount, even under the un-
favorable conditions of a foul sewer and unventilated soil-pipe, was hardly equal to the one thou-
sandth part of that of carbonic acid, and this
amount diffused in twenty-four hours through the
atmosphere of a house is evidently absolutely insig-
nificant and harmless. With a ventilated soil-
pipe the quantity which can pass through the water-
seal was found to be about four times less, probably far less than what would come into our city
houses through the doors and windows from the
ventilating openings in the streets of the public
sewers.

Drs. Carmichael, Pumpelly and Smith, Naegeli, Wernich, Miquel, and others have shown that disease-germs and bacteria generally have the same "mechanical affinity" for water which we observe in all solid particles, particularly of organic nature. They cannot rise spontaneously from the surface of water at rest, and at the normal temperature of our houses. It is only when the surface is violently agitated, or when gaseous bubbles rise to the top and burst, that these particles are released and dissipated in the atmosphere. With a ventilated soil-pipe no such effervescence in the water of a trap can take place, and the agitation of its surface caused by properly arranged flushing does not throw water out of the trap nor allow of the escape of any germs of disease, for any water which may be washed up on the sides of the trap above the normal water-line is quietly carried down again by the upper flushing stream and swept into the sewer.

The experiments of Dr. Carmichael resemble absolute demonstrations and may be accepted as conclusive. He concludes his report as follows: "Water-traps are, therefore, for the purpose for which they are employed, that is, for the exclusion from houses of injurious substances contained in the soil-pipe, perfectly trustworthy. They exclude the soil-pipe atmosphere to such an extent that what escapes through the water is so little in amount and so purified by infiltration as to be perfectly harmless, and they exclude entirely all germs and particles, including, without doubt, the specific germs or contagia of disease, which we have already seen are, so far as known, distinctly particulate."

Tightness of Joints. A fourth principle is that all joints should be permanently tight, and to secure this evident desideratum no material should be used

in jointing which is injuriously affected by any of
the substances brought in contact with them or by
movement produced by changes of temperature,
concussion, or shrinkage.

Soundness of Material. A fifth principle is that
all the material used be sound, and all pipes of even
thickness and capable of resisting a suitable
pressure-test.

Ventilation. A sixth principle is that all the
main lines of soil and drain pipes be thoroughly
ventilated from end to end.

Flushing. A seventh principle is that all parts
of the waste receptacles and pipes be thoroughly
flushed with water from end to end in such a manner as to remove all foul matter instantly from the
house as soon as it is generated.

Automatic Operation. An eighth principle is that
the working of all parts of the plumbing system
should be as far as possible *automatic.*

Noiselessness. A ninth principle is that the
operation of all parts of the work should be noiseless.

Economy and Prevention of Water-waste. Finally,
all parts of the work should be economical in construction and designed in such a manner as to avoid
the chances of waste of water through leakage.

These ten broad principles are not only accepted
by all the leading sanitarians, but are self-evident
and may be at once adopted as axioms without discussion. In the manner of applying them in practice, however, we do not find the same universal
harmony. Where all are in accord I shall make no
reference to authorities. But where there is a difference of opinion among experts I shall call attention to the fact, so that each one present may
form an independent judgment of his own.

The first subject we shall consider will be the trap.

Its form depends upon the nature of the work it is called upon to do ; the form which is suitable for a water-closet being quite unsuitable for other fixtures.

The agencies which tend to destroy the water-seal and efficiency of traps are : siphonage, evaporation, back pressure, capillary attraction, self-siphonage, leakage, and the accumulation of sediment. These agencies must therefore all be considered in the design of our trap. What form shall we give it to enable it successfully to withstand them?

We find that if we adopt the simplest possible form, that of the S trap, which consists merely of a bend in the pipe deep enough to make a seal, we obtain a device which, with proper flushing, is sufficiently self-cleansing and furnishes the easiest outlet for the water. But it is unable to do any more without external aid, and quickly loses its seal under the slightest disturbance of atmospheric pressure produced by a sudden flow of water through the pipes with which it is connected.

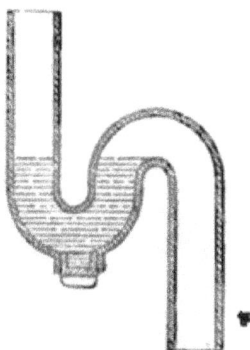

Fig. 2. Ordinary S trap.

Three methods have been employed with a view to preventing the destruction of the seal by siphonage.

One is to ventilate each trap by connecting it with a special ventilating pipe constructed for the purpose.

A *second* is to increase the size of the upcast limb
of the trap until it becomes a " pot," or " reservoir,"
trap large enough to accomplish the same result
without external aid.

A *third* method is to construct the trap in such
a manner as to render it both antisiphonic and self-
cleansing at the same time.

The *first method* adds greatly to the cost and
complication of the work. It has given rise to the
so-called " trap-vent" law, which rigidly requires
every trap, under all circumstances, to be ventilated.

In regard to the practical working of trap ventilation two things have been found : —

First, that it is not always efficient in preventing
siphonage.

Second, that it *is* always more or less active in
destroying the seal through *evaporation.*

Nevertheless, this method still has a few advocates
of recognized ability. But they now adhere to it
chiefly, if not entirely, on account of an alleged
indirect advantage produced by the air-current in
partially oxidizing *foulness in the waste-pipes.*

The *second method* is both inexpensive and
simple and is much more efficient in resisting
siphonic action than the first. It has, however, the
serious disadvantage of involving the use of cesspools or filth-retainers in the house, and such retention is in violation of a leading principle of sanitary drainage which calls for complete removal of
foul matters from the premises the instant they
are generated.

This method has however a very large number of
advocates who consider the retention of a limited
quantity of filth in the trap less of an evil than the
dangers of difficulties coming from trap-venting.

They claim that a guard which is only *sometimes* reliable is worse than none at all as giving a false sense of security, and that the purification of the branch waste-pipes can be effectually accomplished by powerful water-flushing, making the induction of the air-current for this purpose quite superfluous. They find, moreover, that abundant aëration goes on without the aid of the vent-pipe both from diffusion of the air in the ventilated soil-pipe, and from the powerful influx of air induced with, and after, the water-flushing at each usage of the fixture.

The *third method* is the simplest and least expensive of all. It is more reliable than either of the others in resisting siphonic action, and does away with the serious objection of the second method: that of filth retention.

It has already the advocacy of many of the leading sanitarians of the country and promises to be universally adopted as soon as it becomes generally known.

Let us now examine these three methods carefully in detail, since the question is not only one of the most important and interesting ones in the whole domain of sanitary plumbing, but its investigation will throw light on every other part of the subject.

Trap-ventilating. Until very lately it was supposed that trap ventilation afforded a reliable cure for siphonage, and under that supposition the trap-vent law was made. This law has been in operation but a very few months and in a few large cities, yet it has been in force long enough to show in the first place that it is by no means able to do what it pretends to even when the vent is newly and skilfully applied, and in the second place that it gives rise to new evils as great or greater than

those it was intended to obviate, and in **the third**
place that the vent-pipe itself tends to become **foul**
in usage, and that in some **cases** the accumulation
of foulness goes **on to** such an **extent,** especially at
its point of connection **with the trap, as** to completely close the **air-passage and destroy its operation.**

We will first test the efficiency **of the** trap-vent
when it is new and clean and afterward consider the
question **of** its partial or complete closure by **filth**
accumulation.

Tests on Traps Newly Ventilated. In making **these**
tests two **points have been very** carefully followed,
and these must be distinctly **understood.**

In the first place the apparatus and **arrangement**
used is precisely the same in character as is **found**
most commonly in **the best** ordinary **practice.**

A large number **of return bends and** a very long
stack of piping has **been put** together in order to
permit **a** variety of different tests to be made with
a single **compact apparatus.** **But as** openings have
been made **in the** pipe at various points, **we are**
able to cut off **one** or all of the bends and any part of
the length **of pipe** at will. Hence, **the** apparatus
may be made to correspond with that **in** any form
of house we desire to imitate.

In the second place, though **our** tests will be very
severe, they will, nevertheless, **be no** more so than
is often encountered in practice. **Our** object is not
to show what *usually* takes place **but what** at any
time *may* take place.

We do not of **course pretend to say that a** new
vent-pipe **can never protect a** trap, but that **it cannot** always be relied **upon, and** that this being the
case, it **affords a** false sense of security, and is
therefore worse than **nothing, for** we can **never
tell at what moment it will fail in its** duty.

If we are to be forced by the law to put our clients to the great expense and danger of ventilat-

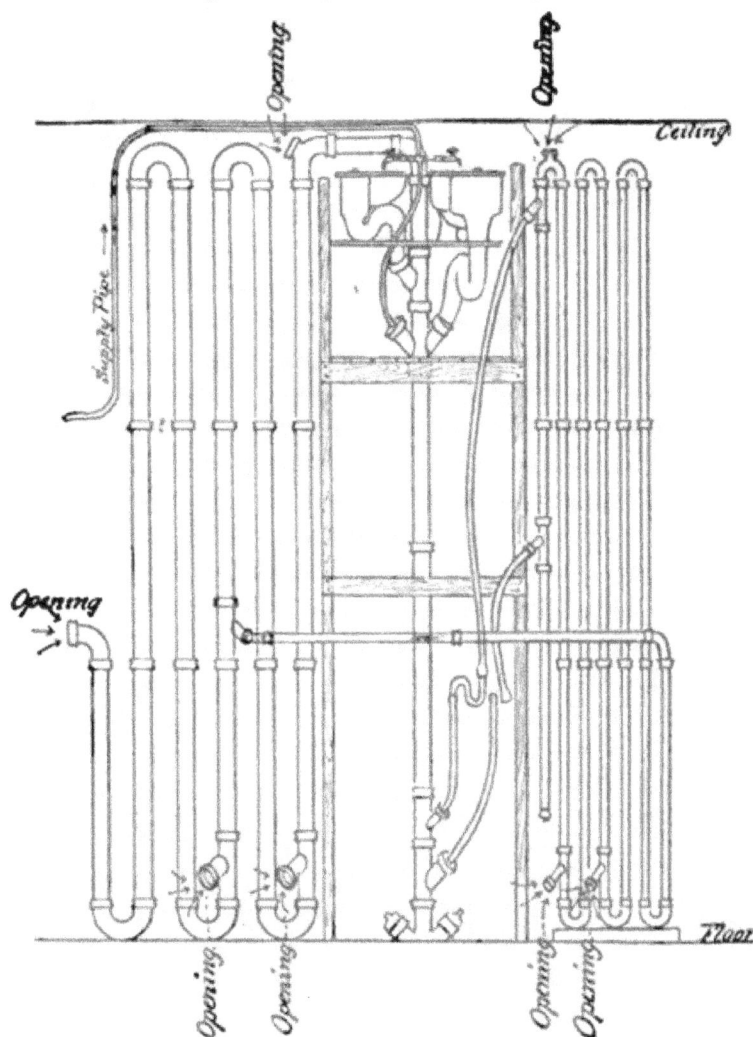

Fig. 3. Apparatus used for trap testing.

ing every trap, we have a right to demand : first, that the means employed shall actually afford us the security it pretends to, and not fail at the first critical moment, and second, that no other simpler and better means exist for securing the desired results.

Our apparatus consists of a stack of four-inch soil-pipe with two ordinary plumbing fixtures ten feet above the wastes of the traps to be tested. On the left is a **Jennings** closet and on the **right** a Zane : kinds which **have** been perhaps until **lately the** most popular in this country.

Fig. 3 represents the apparatus used. The **distance from** the floor to the upper platform **which supports the water-closets is** fourteen and a **half feet.** From the **floor to the** ceiling is seventeen feet.

It is hardly necessary to **explain to the** present **audience that the** smooth **bends and** returns **we have used add but very slightly to the friction.**

Smooth **bends of a** radius **equal to, or greater** than, the diameter **of** the **pipe are found to have very** little **effect in** retarding **the flow of fluids.** To **show** just the effect **such bends produce in the present case we have provided openings at different points in the** length of **the** piping **and after experimenting with all** the **bends, we** will make **other tests without them** or with **only a** portion **of them, and compare the results.** We **have used the** ordinary **two-inch iron pipe for** back ventilation.

Just above the floor of this room **we have provided two Y branches for** the **trap** waste-pipes **to be tested, using ordinary** one **and a** half-inch **bath or** basin **lead** waste-pipe **of the average** length. We **find the length of** these branches **within** reasonable **limits does not appreciably** affect **the siphoning action.**

Experiments on a one and a half-inch S trap. The **first test we will** make on **an** ordinary one **and a half-inch S trap** unventilated. The soil-pipe we **leave full** length. The seal **is two** inches deep.

(*Discharge of Z and J together.*) We see that a single discharge of the two closets has completely destroyed the seal in a second, leaving scarcely a drop of water in the trap.

If we shorten the soil-pipe one half by removing the plug at forty-five feet length, we find substantially the same result. (*Discharge of Z and J together.*) The seal is again instantly destroyed.

This is the simplest possible illustration of the phenomena of trap siphonage, and so far the result is probably familiar to most of the audience here to-night. All are aware that an unventilated lavatory S trap with even an unusually deep seal possesses scarcely any power to resist siphonage. When the falling water in the soil-pipe produces the partial vacuum behind it as it descends, if the soil-pipe extension above it is short and closed at the top, the action is at its maximum because there is very little air to expand. If the pipe is short and open it is at its minimum. If it is long and closed still the action is powerful, but if it is long and open above, a medium effect is produced, and this is the condition we have to-night in our apparatus.

Let us next see what the effect of a discharge of a single closet will be, leaving the soil-pipe forty-five feet long. This cuts off three bends, leaving it four.

(*Discharge of Z alone.*) The Zane closet alone has siphoned out the trap in a single discharge. Let us try the Jennings alone.

(*Discharge of J alone.*) The Jennings alone has also instantly unsealed the trap.

Thus we see that either a Zane or a Jennings plunger-closet is easily able to destroy the seal of an ordinary S trap under the simplest conditions of

plumbing. Any other form of plunger-closet or
any valve or properly constructed hopper-closet
would probably do the same.

Let us now ventilate our trap with a vent-pipe
the full size of the bore of the trap. Leaving the
soil and vent pipes the full length we will discharge
the closets as before.

(*Discharge of Z and J.*) The first discharge has
reduced the seal from **two** inches to one and a half
inch. It will be observed that our vent-pipe is
actually considerably larger than the bore of the
pipe which is contracted at the bends, so that the
protection afforded by this is greater than it would
be in ordinary practice.

Another discharge has lowered the seal to one fourth
of an inch. A third discharge has completely destroyed the seal, leaving an open passageway into
the house for sewer gas.

Thus we see that with the long stack of pipe our
ventilation has signally failed. We will now cut
off half the bends and half the length both of soil
and vent pipe, leaving a medium length of each of
forty-five feet.

A discharge of the closets has lowered the seal
to one and one-half inch. Four discharges have
destroyed the seal.

At the last lecture we found it possible to break
the seal by discharging only one of the closets at a
time. But it required eight repetitions of the discharge to do this, and we shall accordingly omit the
experiment to-night for want of time and because
our subject extends over a wider field than before.
Our next experiment will be with an ordinary one
and one-fourth inch vent-pipe which is really the
size of the trap under consideration. Omitting the
tests with a medium length of soil and vent pipe

which broke the seal of this trap in **two** discharges, and also the test with a single **closet, we** will **shorten** the ventilation-pipe to fifteen feet **by cutting off** the two-inch **iron pipe and all** its bends **alto-gether.** This gives us a shorter vent-pipe than we should ever **be** likely to encounter in **practice.** Hence, if the friction **produced in** this short **length of pipe is enough to prevent the** effectiveness **of the vent,** anything **longer than this would certainly de-stroy** its **power.** The soil-pipe is of medium length.

Discharge of Z and J. Four simultaneous discharges of the closets have destroyed the seal of **our** trap, fully vented with a new in the manner **required** by the law, showing our expensive **venting to be** utterly untrustworthy, **even under the simplest conditions.** In the **experiments made for the City** Board **of** Health the **same results** were obtained by the dis-charge of a single plunger-closet.

Four-inch by four-inch Y. We **have, up to this** time, **used a** four-inch by **two-inch Y branch to** connect **our lead branch** with the main soil-pipe. **In our experiments for** the Board of Health we were **severely criticized by** *The Sanitary Engineer* for using a four-inch by four-inch Y branch, which, we were told, **would produce an action at** least **four times** as **powerful as the smaller branch. In order to** test this **we will connect** our **waste with a four-inch** by four-inch branch immediately below **the one we** have been using, and repeat the last test under **the new** conditions. I **would** caution those **of the audience** who are seated nearest **the** trap **to hold firmly to** their seats, which **have been** tightly screwed to the floor in order to prevent them from being sucked bodily into the drains by the pro-digious siphonage power of **our four** by four-inch branch.

(*Discharge of Z and J.*) We find no appreciable difference in the two Ys, and I think those gentlemen can now safely release their hold upon the furniture.

We have records of comparative tests made with two such Ys, made in exactly the same position on the apparatus, showing a greater rather than a feebler action produced by the smaller Y.

Experiments with a Partially Clogged Vent-pipe. When the mouth of the vent-pipe has been partially closed by gradual deposit of sediment, the supply of air through it is proportionally retarded, and it becomes less and less of a safeguard against siphonage. We have made a great many experiments in this field and found the resistance exactly proportioned to the size of the vent-pipe.

In the tests for the Board of Health we used a straight stack of pipes without any bends. The siphonic action was somewhat more severe in all the tests.

Secondary Office of the Vent-pipe. It remains now to examine the secondary office of the trap vent-pipe, namely, the aëration of the branch waste-pipes, promoting decomposition in them, and carrying off the gaseous products of such decomposition.

Some years ago, before it became customary to ventilate all the main lines of soil and waste pipe, as all sanitary engineers are agreed in recommending now, there accumulated in the upper part of the pipe-system large volumes of dangerous and corrosive gas generated by the decomposition of the heavy deposits in the large soil-pipes throughout their entire length. These gases, never liberated as they are now by a constant current of fresh air passing through the main pipes, sometimes formed in such **large** quantities as to eat through the metal **and** escape **into** the **house.** The water-flushing **from the** feeble pan-closets of **that** time was quite insufficient **to** purify the **main-pipe lines, and seri-ous** difficulties **arose.**

Now, however, the **case is** very different. **All** our main pipes are thoroughly ventilated, **and a far** more liberal flushing **is** occasioned by the **use** of modern **hopper-closets. This** comparatively fresh **air of the soil-pipe** distributes itself by diffusion **through the** branch wastes, and gases can **no** longer **collect** to any harmful degree unless they **are** of unusual length **and** insufficiently flushed.

Consider the case of a **short-branch waste-pipe leading from a well-constructed washbasin and con-nected** with a **well-ventilated soil-pipe.**

Fresh air is constantly passing through the soil-pipe, carrying off the products of combustion as fast as they **are formed. If** the lavatory be fre-quently used **and** properly constructed the short-branch **waste-pipe is scoured** from end to end and **kept very free from foul matter.** Fresh air is dif-**fused easily from the** soil-pipe through this **short** branch as far up as to the **trap. If** the fixture is rarely used **the** last **thin deposit** of soap dries up on the sides of **the** pipe, and **what** little decomposition goes on then is inappreciable, and the products are removed by diffusion, **or, if they** are absorbed in the water of the trap, **what could** escape from its surface would, **as** we have **seen, be** absolutely infin-itesimal and harmless. **Not so if we ventilate this short-**branch waste, as now required by law. A few **days is sufficient** to evaporate **out all** the water from the trap, and soil-pipe air may then enter the house freely. This is no careless assertion founded on theory. It is the result of a series of very careful

experiments made by myself, and published in the
sanitary journals, and it is the experience of experts who have examined the working of the trap-
vent law during the short period since its enforce-
ment.

Consider next the long-branch wastes of lavatories. We will suppose the fixture to be a washbasin
or bathtub used every day. If the outlet be properly constructed the discharge of the fixture will fill
the pipe so full as completely to drive out the air that
was in it and fill it with a volume of perfectly fresh air
from the room. Every one has observed fresh air
being sucked into the outlet of a lavatory, at the time
of discharge, in volume sufficient to renew the air of
the branch waste-pipe many times over, even with
basins improperly constructed as they are.

We will suppose the fixture to be very seldom
used, say not oftener than once a week or month, as
in a spare room. The last charge of water passes
off and the pipe dries up. I believe that what decomposition would then go on in pipes connected
with a properly constructed lavatory would be utterly harmless; and more than this, I believe there is
no case on record of harmful corrosion ever being
found on such branch wastes. It certainly would
not do to ventilate the trap of such a fixture left in
periodical disuse; for evaporation would unseal its
trap in its intervals of rest, and far greater damage
would arise than could come from the unventilated
pipe. Those who do not possess this degree of confidence have only to arrange their fixtures in such a
way as to avoid long-branch wastes, and the difficulty will for them be avoided.

Consider now the question of branch wastes from
kitchen and pantry sinks. Every one knows that
grease and sediment from these fixtures will at once

clog up in time any part of a trap not scoured by the water. We find the upper part of ordinary pot-traps always fouled with grease in such cases. The mouth of the vent-pipe taken from the top of such a trap also becomes similarly clogged, and it is probable that in whatever way the vent-pipe be attached to the trap of a sink it will surely become clogged and inoperative in time.

The only case in which trap ventilation can be recommended, as it seems to me, is in connection with certain kinds of water-closets. The consideration of this branch of my subject must be left for another time.

I find, therefore, no advantage whatever in trap ventilation, with the above possible exception. There are several disadvantages, which, summed up briefly, are these: *First*, it destroys the seal by evaporation when ordinary S traps are used and when the vent-pipe is taken from the crown, as the law in some places requires. I find that if the vent-pipe be taken from some points six inches or more below the crown evaporation does not go on, or it goes on so slowly as to be harmless with traps holding a reasonably large body of water. With S traps, however, it is *necessary* to ventilate at the crown, if they should be used at all, in order to prevent self-siphonage.

Second. The vent-pipe does not accomplish its objects, and hence affords a false sense of security.

Third. It increases the unscoured area of the trap, making it a cesspool. The ventilated S trap is used instead of a reservoir-trap by the advocates of trap ventilation for the sake of avoiding an unscoured chamber. But in doing so they add a sediment chamber, which is not only greater in extent of surface, more easily fouled and less easily cleansed, than that in the pot-trap, but one which is far more dangerous, inasmuch as its fouling, even to a limited extent, involves the destruction of the whole system. This chamber is as certain to become inoperative after more or less use as is any reservoir or cesspool in a trap to become clogged with deposit. It is so placed and of such a form that it must inevitably receive spatterings from the filth-laden wastewater, without benefiting by its scour. I have found, by repeated test, that the water discharged from a washbasin with a large outlet and trap placed a foot or more below is thrown up over ten inches into the vent-flue at every discharge. Thus a very large sediment chamber is formed. The deposit of sediment may be rapid or slow, according to circumstances, in some cases requiring years to reduce the size of the vent-opening to the point of inefficiency. In others this will occur in a few days.

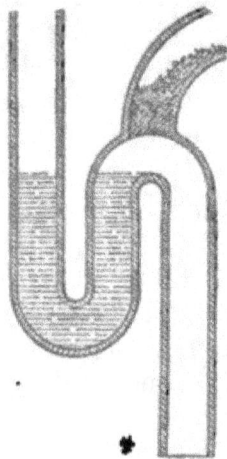

Fig. 4. Trap vent clogged at the mouth.

Fourth. It retards the outflow of the wastewater about thirty-three per cent. This is owing to the friction of the air-current entering with the water during the discharge.

Fifth. It renders the discharge noisy. The same air-suction which delays the water produces a disagreeable roar when the water discharges rapidly.

Sixth. It complicates the plumbing and adds to the danger of leakage through bad jointing and increased material.

Seventh. It aggravates the danger arising from capillary attraction; and, finally,

Eighth. It seriously increases the cost of plumbing, an increase which amounts to as much as from
five to ten per cent. on the total cost of the

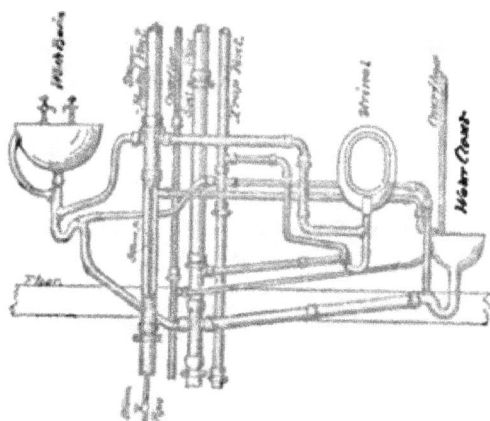

Fig. 5. Complexity with insecurity.

plumbing in new work and indefinitely in old work
in which the trap ventilation sometimes becomes by
far the greatest part of the work to be done.

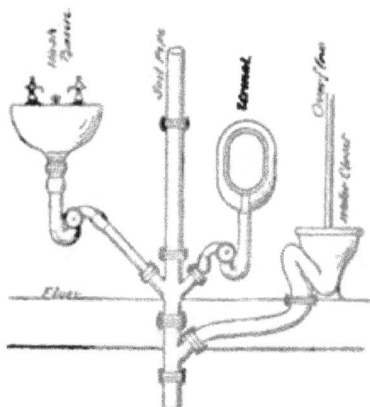

Fig. 6. Simplicity with security.

Figs. 5 and 6 represent three fixtures plumbed
by the two different methods, the first with, and the
second without, trap ventilation. In the first draw-

ing the overflow-passage and the house-side of the trap are ventilated as well as the sewer-side, and the loss of the water-seal through evaporation is very rapid. This double trap ventilation is not common, but yet is occasionally carried into execution by some of our more radical enthusiasts for branch-waste venting.

In the second drawing "Sanitas" traps are used which require no ventilation to prevent siphonage.

A washbasin, having an outlet large enough to fill the waste-pipe and trap "full-bore," scours them and keeps them free from deposit.

The use of a urinal is rarely to be recommended. It is only introduced here for purposes of illustration.

Second Method. Let us now examine the second method of obtaining security against siphonage. This consists in the use of a large unventilated pot, or reservoir, trap. A small pot-trap will not resist siphonic action, but a large one will. Their power of resistance is exactly in proportion to their size. Nothing smaller than an eight-inch pot-trap, which I have here, can be relied upon in all cases. A six-inch pot-trap will sometimes be siphoned out by discharges occurring in common practice. A five-inch pot-trap siphons out much easier. An ordinary four-inch trap has very little resisting power unless its seal is unusually deep. Three-inch and two-inch traps are altogether useless.

We will test practically the action of siphonage on a four-inch pot-trap of the usual depth of seal.

(*Discharge of Z and J.*) We see that three discharges are sufficient to break the seal of this trap. In our last lecture we found that either the Zane or the Jennings closet alone was able to destroy the seal in eight and six discharges respectively.

Thus we see that only the largest sizes of pot-traps are reliable. To be secure in all cases, if we use pot-traps, we are required to have them as much as six or eight inches in diameter, and constantly inspect them to see that they are free from deposit. Traps of this size are veritable cesspools and as such are to be avoided wherever it is possible. They are, moreover, expensive. A plumber's scale of charges for these traps is at the rate of one dollar for every inch in the diameter of the trap. Thus a five-inch, six-inch, and eight-inch pot-trap costs $5, $6, and $8 respectively.

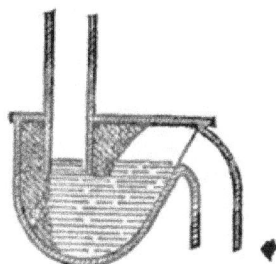

Fig. 7.　D trap clogged.　　　　Fig. 8.　Bottle-trap.

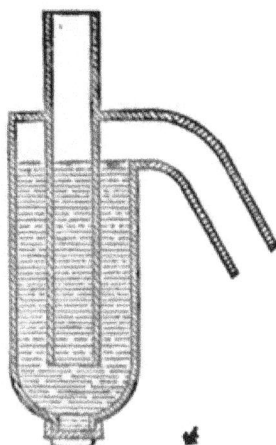

The pot-trap is, morever, bulky and unscientific in construction. Its cleanout cap is faultily arranged at the top, where, if improperly adjusted, it will allow the escape of sewer gas without warning. The cleanout cap of a trap should always be wholly or in part below the normal level of the standing water in order that if an unsound joint occur it will at once be detected by an escape of water and the defect remedied. It is better to endanger the floors or plastering than the life or health of the owner.

To ensure tightness the plumber is obliged to screw the cap on so hard that the house-owner is rarely able to unscrew it for examination or cleansing. Hence the plumber has to be **sent** for. Illfeeling is aroused and the plumber is referred **to** in terms often lacking in refinement and politeness.

Under the name " reservoir " traps I include all water-seal traps which are not self-scouring. It **includes the** old-fashioned **D trap** (Fig. 7), **the Globe trap, and the Bottle trap (Fig.** 8).

Of all the reservoir-traps, the **common pot-trap,** bad **as it is, is the best, as being the** simplest.

Balls, **valves, and gates in traps** add **little or** nothing to their power **of** resisting siphonage, and have no longer any value **now** that it is customary to ventilate the drain **and** soil pipes. They serve **only** as encumbrances and filth collectors.

We come finally to the *third* method of obtaining security against sewer gas, **of** which Fig. 6 forms **the** general illustration. **It** is to give the trap **such** a form **as** to render **it** antisiphonic and selfcleansing at the same time.

ANTI-SIPHONIC **TRAPS.**

Let us first examine the action **of** fluids **in traps** when they are subjected **to** siphoning action **and see** if it is possible to construct a **trap** in such a **manner** as to accomplish these results. To better study the movement **of** the fluids we have had a number **of** S and pot traps constructed wholly of glass.

We must make **use of the** natural forces at our command, the superior gravity and adhesive force of water over air, and construct our trap with **reference to the** laws governing these forces in the movement **of the two** fluids.

Examining first our pot-trap. Under a powerful siphonage air is driven through the water in the body of the trap in the manner shown in this drawing (Fig. 9). A quantity of water is projected out of the trap in advance of the air-column, as shown by the arrows. If the action were continued long enough all the water above the inlet-mouth, even in

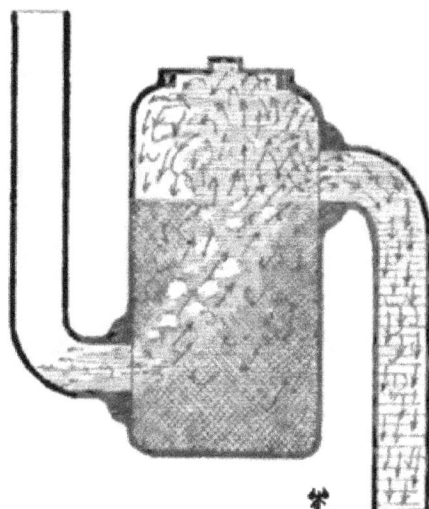

Fig. 9. Movement of fluids in a pot-trap.

the largest pot-traps, would be expelled. It will be observed that part of the water is forcibly thrown up against the top of the body of the trap, whence it is deflected back in the form of spray in all directions. Part of the spray, however, falls across the outlet-mouth, and is sucked out. One of the principal reasons why the S trap is so easily siphoned out is that the curve at the top conducts the water directly into the outlet. Some form of reflecting surface should be used to throw the water back into the trap, and let the lighter air escape to supply the vacuum in the soil-pipe. Such a reflecting surface is found in the flat top of the pot-trap above the outlet-mouth. We will therefore retain this useful feature, but reject the objectionable one of the excess of sectional area in the body over that of the inlet and outlet arms, and we have our first modification, as shown in Fig. 10. The reflecting surface, however, should not be arranged as here shown. The pocket increases the unscoured area of the trap. It is true it is no worse than the mouth of a ventilating pipe, which under the present law it is customary to put at this place. But it is just as certain that such a pocket will become clogged in time as it is that grease and filth will deposit a sediment on everything with which it comes in contact. The higher or deeper

Fig. 10. First modification. the pocket the more readily will the deposit be formed. A shallow pocket might be partially scoured by the force of the water projected upward against it by momentum. In this case a certain portion of each deposit of filth would be washed off by friction and the process of clogging would be somewhat retarded. But let the pocket be deep enough and there will then be parts which will be within the *reach* of the waste-water, but beyond its *scouring effect*. The spray thrown up by momentum will at this height have lost its power. The drops of dirty water will simply rise to their turning-point, deposit their filth, and trickle back again into the trap. The ventilating outlet forms exactly such a pocket. At a certain height

above the crown of the trap the inner surface of this flue will receive the *spatterings* of the filth-laden waste-water, but never receive its *scour*. Hence the area of the vent-opening must infallibly continue to decrease in size more or less quickly, according to the usage of the fixture, until the opening is too contracted to be of any value in resisting the action of siphonage on the water-seal. Moreover, the cool ventilating draught helps to congeal the fatty vapors arising from hot waste-water in the trap and hastens clogging. We will, therefore, simply retain the reflecting surface but reject the pocket. Furthermore, we will slightly contract the inlet and outlet mouths at their junction with the body. This allows the air rushing through the body of the trap to pass *through* the water instead of driving it out before it. A very slight contraction is sufficient. These two modifications make the second step in our improvement, and are shown in Fig. 11. A trap was constructed in this manner, and proved to be very much stronger in resisting siphonic action than an S trap of equal depth of seal.

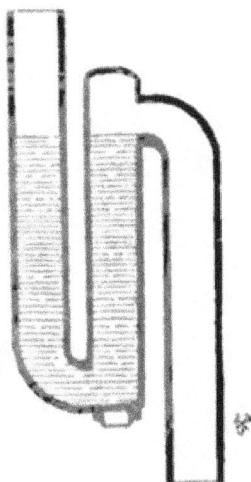

Fig. 11. Second step. Partial contraction of inlet and outlet mouths and reduction of pocket.

Still our trap is very far from antisiphonic. Referring to our glass pot-trap, we shall see that the water projected violently upward from the surface, by the air-bubbles rushing through the standing water under the influence of siphonage, is obliged to pass *twice* by the mouth of the outlet-pipe, once before and once after reflection against the top, and that it is at these moments that it is sucked out and lost. That part of the spray which happens to be thrown farthest from the mouth of the outlet-pipe will be seen to fall back safely into the trap; but that which passes near this outlet, either in rising or after reflection, is drawn out by the concentrated and powerful suction at this point and wasted. And we find that one of the principal reasons why a large pot-trap resists siphonage longer than a small one is that in the large trap the spray has more space above the surface of the standing water than in the small one, so that a smaller proportion of the water thrown up by the rushing air-bubbles passes within the influence of the suction at the outlet-pipe. If our reflecting surface could be placed below instead of beyond the mouth of the outlet this loss could be avoided. Our next step

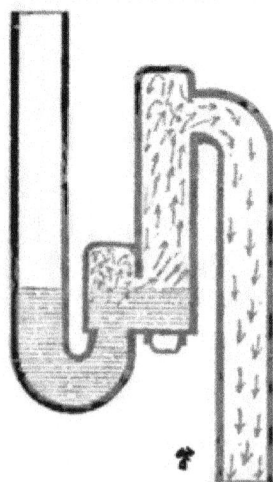

Fig. 12. Third step.

must, therefore, consist in so placing the reflecting surface. In Fig. 12 this has been accomplished, but in an awkward manner. Before this surface can come into service the level of the water must evidently be reduced to the level shown by the shading in the figure. Hence the perpendicular part of the body of the trap above the lower reflecting surface is not placed to advantage. Nevertheless, this trap will resist a very powerful siphonic action, even as it is. The two reflecting surfaces, the lower and the upper, are so effective that this form of the trap has proved more tenacious of its last inch or two of seal than a four-inch pot-trap, although its diameter is nowhere greater than that of the outlet and inlet pipes.

In this and in the preceding forms the depth of seal is too great to allow of a free and rapid discharge of the wastes. The air, in passing through the trap, disturbs nearly all the water in it. Our next step will therefore be to diminish the height of the water-column through which the air has to pass, and thus reduce the disturbance of the water without lessening its volume. It may be done by laying the body of the trap horizontal instead of perpendicular, as shown in Fig. 13. This immediately

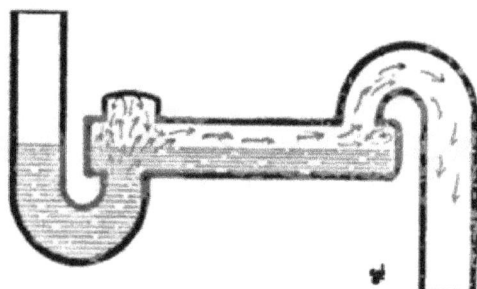

Fig. 13. Fourth step.

gives us a very important improvement in resisting power. The area of the trap is no greater than that in Fig. 11, but it is found to offer double the resistance to siphonage. Moreover, while the volume of water is the same as in Fig. 10, the seal is not so deep. Hence the flow of water through this trap is more rapid than in the former, and its scouring effect correspondingly increased. As soon as the water in this trap has been lowered to the point indicated in the drawing, ample space is left above it for the passage of the air. It is evident that a much smaller body of water is disturbed by the passage of the air than is the case with the trap shown in Fig. 11.

Nevertheless, the trap thus made is not yet sufficiently antisiphonic. It is, moreover, awkward in form and difficult to set in such a manner that it shall remain firm in place. The long horizontal body is liable to sag and lose its form. Moreover, a single reflecting surface is insufficient to separate the water entirely from the air, and a strong and long-continued siphonic action destroys its seal. Other improvements are evidently necessary.

A fifth step consists in increasing the number of reflecting surfaces, and in breaking up the long horizontal body by making it return upon itself in a quadrangle, as shown in perspective in Fig. 14. In this form of the trap we have still further greatly increased the reflecting surfaces and the power of resistance to siphonic action, and we are now able to dispense with reflecting pockets, but we have obtained a trap exceedingly difficult to manufacture, awkward in appearance, and troublesome to clean out in case of accident, as when a match or any such foreign substance is dropped into the waste-pipe and becomes lodged in a bend of the trap. This form of trap must be simplified so as to render it practical, without losing any of the advantages we have thus far arrrived at. Figs. 15 to 19 show the manner in which this may be done, and the arrangement forms the final step of our improvement. We have here retained all the reflecting surfaces ; the horizontal body, which allows the air to pass above the water after a small quantity has been driven out, without disturbing the rest ; and the slight contraction of the inlet and out-

Fig. 14.　Fifth step.

let pipes at their junction with the body of the trap. We have added a cylindrical cleanout cap of glass, and obtained an apparatus which can be readily cast in lead in moulds of iron. The quad-rangular shape of the horizontal body is retained, but the two parallel cylinders are brought to-gether and merged into a single cylinder having a central partition about two thirds of its length, or extending from one end to the edge of the cleanout cap, which at the other end forms about one third of the total length of the cylinder.

Fig. 15. Sixth step. "Sanitas" trap.

In ordinary use the waste-water passes through this trap in such a manner as to act to the best advantage in scouring it. The partition wall in the centre of the body causes the water to scour each side in succession. Thus while in outward appear-ance the body resembles a small pot-trap placed horizontally it has in principle the self-scouring form of the S trap. It must be understood, how-ever, that like the S trap it is only self-scouring when properly set, namely : with a free outlet from the bowl somewhat larger than the inlet-arm of the trap at its largest part, or at its point of junction with the fixture. If set under a fixture giving a clear water-way of only an inch or of half an inch, this trap will not scour itself, nor will the waste-pipes with which the trap is connected. A good-sized washbasin holds, up to its overflow, about two gallons of water. This will escape through an average length of one and one-half inch waste-pipe, running full-bore and having a good fall, in about three seconds. Hence, through such a pipe

the water rushes at a rate of more than half a gallon
a second and fully scours the pipes. With lavatories constructed on this principle, the argument
for trap ventilation based on the supposition that
it is necessary to keep the branch wastes clean
no longer holds good.

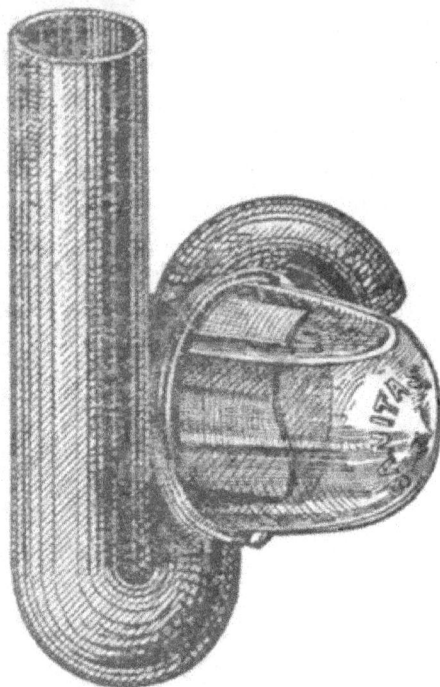

Fig. 16. "Sanitas" trap.[2]

Let us examine now the action of the air and
water in our trap under the influence of a very
powerful siphonic action. We will suppose the
trap to be placed in position under a fixture with
the water standing in its normal condition up to the
level of the outflow, as shown in Fig. 16. When,

[2] This trap is manufactured by the "Sanitas" Manufacturing Co.,
No. 4 Pemberton Square, Boston, Mass.

through siphonic action, a partial vacuum is created
in the waste-pipe below the trap, the water in the
inlet-arm of the trap descends under the influence
of the atmospheric pressure on its surface tending
to restore the equilibrium, until it reaches the dip

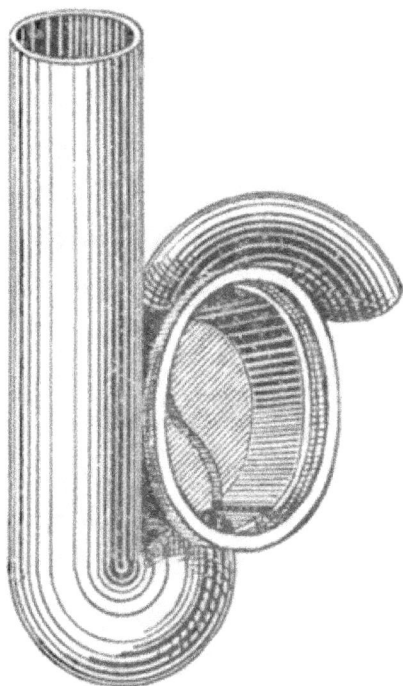

Fig. 17.   Body of trap.

of the trap.   The air then being lighter than water
passes into and through the body of the trap and
drives a portion of the water, not already driven
out, before it into the waste-pipe.   The water remaining in the body falls back and maintains
the seal.   Subsequent siphonic action cannot remove this water for the following reasons : The
water standing in the inlet-arm after its partial
removal from the body of the trap by siphonic

action, as described, is again lowered by a repetition
of the action to the dip. Air again rushes into the
body to fill the partial vacuum and passes into and
through the water standing therein. This water,
though increased in depth by that which enters
from the inlet-pipe, is, nevertheless, shallow enough

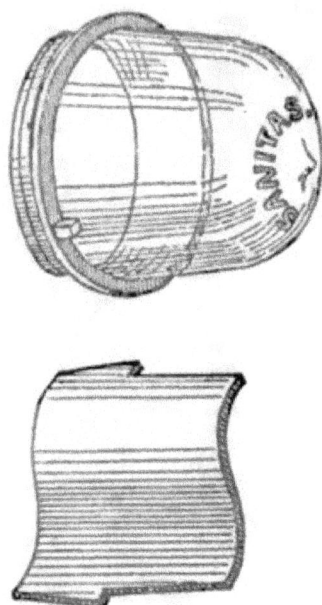

Figs. 18 and 19. Glass cap and bridge.

to give room for its passage. It projects upward
a certain quantity of water in its passage, with
greater or less violence, according to the strength
of the siphonic action produced. This water strikes
the under surface of the partition in the body, and
is partly reflected backward by it, and partly follows
the air-current toward the opening between the end
of the partition and the cleanout cap. Owing to
the greater weight and momentum of the water
over that of the air, the water is reflected back,

while the air passes on. A second reflection takes
place against the surface of the cleanout cap.
More water is thrown back, and a small remaining
portion only succeeds in following the air into the
passage above the partition. Of this small portion
part again is reflected back by the upper inner
curved surface of the horizontal body, and under
very strong siphonic action a few drops may reach
the last reflecting surface at the end of the body
opposite the cleanout cap, whence it is once more
arrested, and the air alone escapes into the waste-
pipe. The spray falling upon the partition and upon
the various reflecting surfaces collects at the bottom
of the body and increases the depth of the seal.
The greater cohesive and attractive force of the
particles of water over that of air aids in the sepa-
ration, since it causes a quantity of the former to
adhere to the reflecting surfaces while the air
escapes. This arrangement of the reflecting surfaces evidently also prevents loss of the water-seal
by the momentum of the water descending from the
fixture.

Although the seal is not excessively deep, yet the
trap, owing to the considerable horizontal extension
of its passages, contains a large enough body of
water to protect it from the dangers of evaporation
and back pressure. The contraction of the inlet
and outlet arms at their junction with the body of
the trap renders it secure against self-siphonage.
The form also renders loss of seal through capillary
attraction impossible, as will be hereafter shown.

When used where trap ventilation is prescribed by
law, this trap can, of course, be ventilated like any
other. The vent may be applied at any part of the
outgo, either at or below the crown. But since,
unlike S traps, its seal cannot be destroyed by self-

siphonage or momentum, the vent need not **be**
applied at the crown. **It may be** applied anywhere
below the crown **far enough away** to quite avoid the
injurious effects of evaporation. Hence, ventilation does not produce the destruction **to the seal**
that it does with other self-cleansing **traps, and may**
be used with **impunity.** Trap ventilation **is, never-**
theless, in this case, as in most others, absolutely
useless, and **its** installation is a total loss to **the**
house-owner.

Having **now** explained the theory of the construc-
tion of the " Sanitas " **trap,** let us make a practical
trial of **its** operation.

The first discharge **of** both closets, the **soil-pipe
extension** being forty-five feet, will lower **the seal**
considerably, say to **a** point below the centre of the
glass, but subsequent discharges will have very little
further effect upon **it,** and when **the** seal has been
reduced to about an **inch** and a half, or in the very
severest possible cases long repeated, possibly to an
inch and a quarter or eighth, even the most **powerful** suction **that** can be applied with an apparatus
used **in** practice will have **no** further appreciable
effect upon it, even though the siphonic action be
strong enough to destroy the seal of a fully ventilated S trap or of a six-inch pot-trap. Ordinary
siphonic action will simply **lower the** water in the
trap enough **to** permit the **passage** of air above
it, leaving a **seal** of two inches **or** more permanently
in the trap. The test which we are about to apply
to **this** trap **is severe** enough **to** siphon **out** completely in a single second **a fully** ventilated S trap
or a four-inch pot-trap.

(*Discharge Z and J.*) After this very severe test
we find a seal left of one and one-half inch.

(*Repeated.*) Only an *eighth* of an inch has been
removed **by** a second discharge.

(*Repeated ten times or more.*) Five repetitions of
the discharges have lowered the **seal less** than an
eighth of an inch, leaving a full **seal of** one **and**
one-fourth inch. Five **further** repetitions **pro-**
duce no further visible effect on the seal. **In our**
previous experiments **we have** repeated **the test**
fifty times without **apparent** diminution **of the seal.**

Snow. **We** will **now apply a test stronger than**
any we have tried this evening. It is **a test severe**
enough, as we found in making **the experiments for**
the City Board of Health, to siphon **out a pot-trap**
eight inches in diameter.

We will retain the full length of **our** soil-pipe and
stop up the opening above **the roof** with oakum.
Then, by discharging both closets together, we shall
produce a suction as great as any which could **pos-**
sibly be produced in practice **as when the top of the**
soil-pipe is closed up by ice or snow.

The first discharge has left one **and one-half**
inch of **seal.** At the end **of five** discharges **there**
is still one and one-eighth inch **of** seal left and
five more produce **no** further **apparent** diminution
of it. Thus we see the **seal of this trap cannot be**
broken by any siphonic **action we can try with or-**
dinary apparatus used **in plumbing.**

Self-cleansing Property of the " Sanitas" Trap.
It **remains** now to determine **if the " Sanitas " trap**
is actually **as self-scouring as it is** claimed **to be.**

We have had **a "** Sanitas **"** basin and trap set
above a waste-pipe of glass in order to examine the
scouring action of the water discharged from **a**
basin with a properly proportioned outlet **both on**
the trap and the pipe below it.

(*Discharge of the Basin.*) We see that the water
rushes through the trap and waste-pipe at a very
rapid rate. The basin, when filled to the brim, holds

about two and one-half gallons. It empties itself,
when set with a waste-pipe having a good fall, in
about three seconds. Hence, the water flows at the
rate of nearly a gallon a second, and has an enormous scouring-force on all the branch piping connected with it.

We will now throw various substances into the
bowl and trap and see whether they are retained in
them or not.

We will first try a quantity of coarse coal ashes,
and, to make the test somewhat severe, we will remove the strainer and throw into the trap pieces of
coal with the ashes nearly an inch in diameter.
Filling now the basin we find a single discharge has
removed all the dirt and a second discharge has left
the trap and waste-pipe as bright as ever.

I will now form a paste of softsoap and loam.
The loam is a mixture of earth and clay. The
combination of this and the soap forms in large
quantity the kind of waste matter to which washbasins are most accustomed.

All this matter is instantly carried through the
trap and waste-pipe, and, after a second flushing,
it does not leave a stain behind.

After trying a few other substances we will make
a strong solution of soap and dirty water and let it
dry on the pipe and then see if it will wash off
after drying.

Large pieces of hair-felt, strings of jute and
tow, coarse gravel, pieces of stone an inch in
diameter, nails, and matches are all whisked
through the trap as easily as if they were nothing
but house-flies. In short, every kind of substance
likely to be met in usage, and a great many others,
are carried through and away with speed and cer-
tainty, and the self-cleansing power of the trap is
demonstrated.

We have found the soap dried on the glass tube is completely removed by the strong flushing from the basin. (This last test was tried on another occasion before the lecture.)

Capillary Attraction. The seal of a trap is sometimes slowly and silently drained off by bits of hair, sponge, or twine which get caught across the outlet of the trap, as shown in Fig. 20, and draws out its

Fig. 20. Seal of S trap destroyed by capillary attraction.

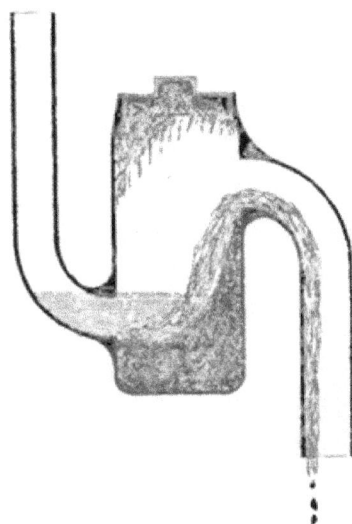

Fig. 21. Seal of pot-trap destroyed by capillary attraction.

water by capillary attraction. Numerous experiments have been made of late on this insidious enemy to the life of water-traps, and it has been found that there is a limit to the height which these substances will carry the water above its normal level. We find this limit of height to be within three inches for small quantities of long and fibrous substances such as might get lodged in traps. We must, therefore, form our trap in such a manner

that the water will have to travel along the fibrous
substance more than three inches before its seal
can be broken. The "Sanitas" trap has been
so constructed (Figs. 22 and 23), and in no

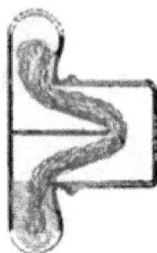

Fig. 22. "Sanitas" trap resisting capillary attraction.

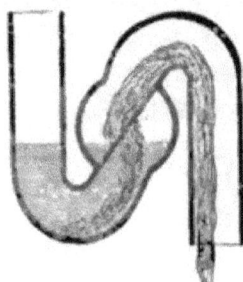

Fig. 23. "Sanitas" trap resisting capillary attraction.

case has it been possible to destroy its seal by
the capillary attraction of substances which could
be lodged in it in practice.

Back Pressure. Back pressure is a force now but
little to be feared in plumbing. Before it became customary to ventilate our waste and soil pipes, pressure
in the sewers, either from winds or tides, or the heat
of steam or chemical action, sometimes produced a
serious back pressure in our house-pipes. Now we
no longer encounter the difficulty from these causes,
since we are accustomed to have our pipes properly
ventilated. It is only under certain rare conditions,
such as when a trap is situated near the bottom of a
tall stack of pipe and close to a sudden bend, that
back pressure is produced by falling water compressing the air in advance of it. The bend in the
soil-pipe prevents the escape of the air below as
fast as it accumulates above under the falling water-
plug.

To resist this pressure it is only necessary to have
a sufficient body of water in the trap and to set

the trap at a distance below the fixture it serves
sufficient for this water to form in the pipe when
subjected to back pressure, a column from twelve to
sixteen inches long. (Fig. 25.) The weight of
such a column is ample to withstand any back pressure ever now encountered in good plumbing.

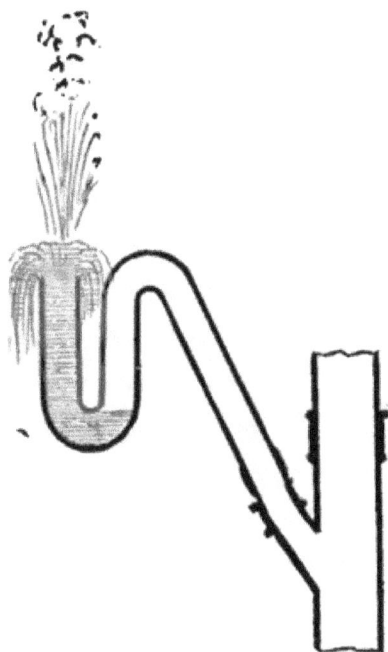

Fig. 24.   S trap emptied by back pressure.

The "Sanitas" trap is made to contain a body of
water heavy enough to easily resist any back pressure it can ever be called upon to bear in modern
plumbing work.

Evaporation. When traps are not ventilated
evaporation goes on with such slowness as to be
scarcely perceptible. Nevertheless, it is best to
have the trap contain as large a body of water as is

consistent with its self-cleansing properties.   An
ordinary one and a half-inch S trap holds about
three eighths of a pint of water.   A one and a
quarter-inch S trap holds less than one fourth of a
pint.   The " Sanitas " holds one and a half pint,

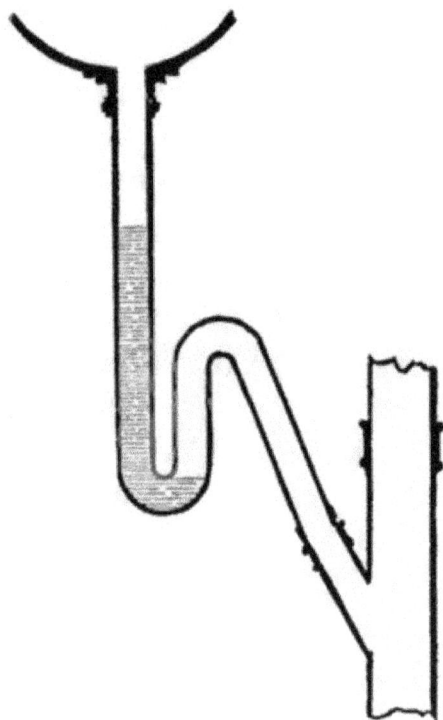

Fig. 25.   Deep seal S trap resisting back pressure.

or about as much as an ordinary three-inch pot-trap.
This is sufficient to last, under ordinary conditions,
over a year without renewal when the trap is unventilated.

Where the trap is ventilated, however, in the
manner customary under our present plumbing
laws, the seal of an ordinary machine-made S trap is

licked up by the air-current in a very short time, varying in my own experiments from four to eleven days.

Size and Material. Traps for the smaller fixtures should be manufactured in one size, that is, of a capacity sufficient to fill the usual one and a half or one and a quarter inch waste-pipe fullbore. In other words, the size of the traps should be governed by the size of branch waste-pipes. These pipes should never exceed one and a half inch in diameter, except for water-closets. Waste-pipes should not be less than one and a quarter inch in diameter. Hence, the capacity of the trap should not be less than this at any part, and to be self-scouring should not exceed this capacity at any part. The cleanout cups should be made of glass or of metal. Glass should be used for washbasins only, and then only when a possible fracture will not produce serious damage to frescoed ceilings below. With bathtubs, sinks, laundry-tubs, and all other fixtures metal cups should always be used, since even the best annealed glass is liable to be broken by sudden changes of temperature or by careless usage.

PART II.

Fig. 25. Ordinary basin-strainer.

THE character of our lavatories is a matter of very much greater importance than is usually supposed. We have been in the habit of selecting our washbasins and bathtubs purely from a standpoint of convenience, appearance, and economy. Sanitary considerations have been quite overlooked, in the belief that they have little or nothing to do with the form of these particular fixtures, so long as their traps and waste-pipes were properly made.

This is a very serious error, and particularly so in relation to washbasins, in the choice of which sanitary considerations should outweigh all others. We say this advisedly, and for the following reasons: As usually constructed, the outlet is altogether too small in proportion to the size of the trap and waste-pipe. The result is imperfect flushing of these pipes, gradual accumulation of filth in them, and the various serious evils to which such accumulations give rise. Fig. 25 shows the actual dimensions of the ordinary basin-strainer. It will be found, by accurately measuring these figures, that the amount of water-way is just equivalent to that of a three-fourths inch pipe. A very short usage soon reduces this meagre opening, through the

collection of sediment and lint, to a still smaller
stream. Accordingly we find that by far the greater
part of the ordinary basins now in use discharge
a stream not over half an inch in diameter. The
waste-pipes are usually an inch and a quarter or an
inch and a half in diameter, a capacity which is
given for the purpose of ensuring the safe removal
of the water delivered by two supply-faucets running full force, under medium or high city pressure,
and escaping through the outlet and overflow passages combined, together with a possible simultaneous discharge of the adjoining fixtures entering the
same waste. Now a half-inch stream of waste-
water, trickling through pipes capable of delivering
ten times as much, fouls, but does not scour, them.
I have taken out such pipes and found them more
than half filled with slime and filth, and in places
where the pipe ran nearly horizontal, or made sharp
bends, I have found them nearly filled with the
putrefying mass. No amount of ventilation can
cleanse such pipes. But the sediment was soft and
gelatinous, and would easily have been swept away
by the powerful discharge of a basin filling the
pipes full-bore.

Besides the important sanitary advantage of a
rapid discharge, we have others of economy and
convenience. To empty an ordinary basin requires
a very considerable amount of time and more
patience than the majority of people possess. The
result is that people fall into the habit of washing
from the faucet rather than from the basin, and a
great waste of water is involved. A quick waste
and convenient method of operating and controlling
it results in a saving of water and very great convenience in usage. A knowledge that a sudden discharge of a basinful of water through the pipes acts

as an important sanitary measure, after the manner
of a flushing tank, in cleansing them from end to
end, leads to a legitimate use of the basin and an
economy of water, a consideration which the water-
companies and the public in time of drought will
not be slow to appreciate.

WASHBASINS HAVING CONCEALED OVERFLOW-PASSAGES.

This class of fixture violates one of the first conditions of sanitary plumbing. A portion of the
apparatus intended to carry off waste-water at the
irregular and uncertain intervals of overflowing
becomes fouled without the chance of cleansing
through flushing action, and is placed in such a
position that it cannot be seen or reached without
disconnecting the whole fixture.

Our first subdivision of this class is the ordinary

PLUG-AND-CHAIN OUTLET-BASIN.

We see here (Fig. 26) the concealed overflow-pipe
constructed of lead and so placed as to be altogether
inaccessible. Being
above in open com-
munication with the
air of the room, it
taints it with the de-
composing soap and
filth with which the
sides soon become
coated. The ordinary
washbasin has no
proper flange for con-
nection with the lead
overflow-pipe; the
joint has therefore to

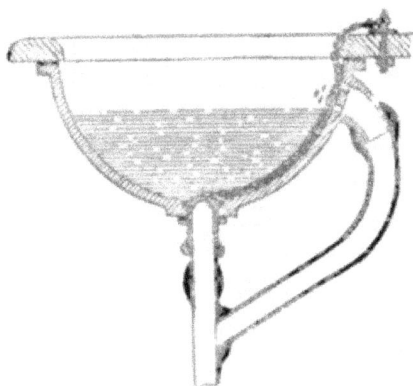

Fig. 26. Ordinary plug-and-chain outlet-basin.

be made with ordinary putty, which **can never be**
made permanently and surely tight. The lead pipe
must be connected with the main waste-pipes above
the trap, and the joint here must be wiped with sol-
der. Thus, to set an ordinary washbasin, the plumber
has two extra joints to make, which **add** both to
the expense **of** the work and to the **chances of** imperfection and leakage. It is an exceedingly **common** thing **to** find the overflow-pipe wrongly **con-**
nected; it is sometimes entered below the trap,
sometimes attached directly **to** the trap-vent, and
sometimes connected **with** the wastes of other fixtures in such a way as **to open, through the** vent-
pipes, an indirect **avenue into the house for sewer-**
gas. It forms, in short, an **unnecessary** and dangerous complication **to** the **plumbing,** and these
basins should never be used.

Many house-owners **stop up the** holes in the
earthenware leading **into the** overflow-pipe **at con-**
siderable inconvenience **to** themselves, in **the hope
of** avoiding the chance of the entrance of offensive
or injurious gases into the house through this chan-
nel. With defective traps, or with traps whose
seal is liable to **be** quickly destroyed by evaporation, siphonage, or other cause, this precaution
against danger would not be useless if the overflow-
pipe connections could be **made** certainly tight, especially when the fixture is left for some time
unused. As they are made, however, it is probable
that **no** such precaution would form any reliable
security.

The use of the plug and chain, which charac-
terizes this type of basin, is another serious defect.
The chain, lying **in every** successive formation of
dirty water, collects gradually in the recesses of **its**
links an unknown quantity **and** variety of filth,

which cannot be entirely removed on account of its irregular form, without the use of special acids or constant scrubbing with a brush, a process never applied to it. The length of wire used in an ordinary basin-chain averages six feet, and has a surface of about fourteen square inches, a surface which, in consideration of the peculiar adaptability of the form of the links for retaining dirt, presents a very formidable area of pollution. To those persons who use their reasoning powers in these matters the idea of washing the face in water defiled by a chain transferred immediately from the dirty water of some unknown predecessor is with good reason exceedingly repulsive. The chain, moreover, frequently breaks, and then the hand must be plunged into dirty water to remove the plug.

The position of the chain and plug at the bottom of the bowl is, moreover, peculiarly inconvenient, inasmuch as they are in the way of the hands, which should meet a smooth, unbroken surface of earthenware rather than the hard and irregular outlines of the brasswork. If this latter consideration appear to some trivial, it does so only because habit has rendered us callous to such defects ; the defect none the less exists, and acquires importance through the frequency of its repetition and the constant use of the fixture in which it occurs. The fact that it is altogether unnecessary is a sufficient reason for its abolition.

CONCEALED WASTE-VALVE BASIN.

Fig. 27 represents a basin fitted with the so-called " Boston Waste," which is very popular. There is probably no form of basin-fitting more faulty in principle than this. It contains two independent, inaccessible, and invisible foul-water passages, one forming the overflow-passage, and the other the

outlet passageway between the strainer and the
waste-cock. This latter passage forms an elongated
cesspool for the defilement of the clean water entering the basin. After using the fixture, the waste-
water escaping through this channel deposits part
of its dirt, particularly floating matter and soap-
suds, all along its sides, and leaves it there to be
taken up and applied in a diluted solution to the
hands and face of the next comer. Six wiped

Fig. 27. Concealed waste-valve basin.

solder joints, one putty joint, and five threaded
joints, making twelve in all, are required to adjust
the waste-pipes of this apparatus and its trap below the basin-slab! No wonder the plumber is constantly in requisition to keep in repair such a complicated machine so long as the owner allows it to
remain in his house. Not the least of its defects is
that the passageway for the waste-water through
the ground-cock is usually so small (scarcely a quarter of an inch wide) that a small deposit of sediment will entirely prevent the outflow of the water.
The "Boston Waste" cannot be too strongly condemned. The great extent of its use in spite of its

high cost shows how little knowledge the **public**
have in these matters, and how important **it is that**
their attention should be called to them.

THE STAND-PIPE OVERFLOW-BASIN.

It is not sufficient that **every** part of **our** apparatus should be visible **and** accessible **from** without,
and devoid **of all fouling chambers and** corners, **but**
it is above **all necessary that,** combined with the
utmost **convenience and simplicity** of action **and**
economy **of construction, it should** be so formed as
to **ensure** the complete automatic scouring **of** its
waste-pipes and trap without detriment to the water-
seal of the latter.

A suitable enlargement of the basin-outlet is all
that is necessary to **prod**uce the requisite scouring
action ; but the force of the out-flowing water-column **is** so great **when the** pipes are charged full-
bore **that** it **will siphon out** and completely destroy
the water-seal of an ordinary S **trap** unless it be
fully ventilated at or very near the crown, and **it**
will dangerously lower it even then. This is the
action which we have **called " self-siphonage." No**
injurious effect is produced on **the seal of an anti-**
siphon **trap** by self-siphonage, **but the water** is somewhat reduced in **the** trap below **its** normal level. It
is therefore desirable, and when S traps must be
used extremely important, that such a basin should
be so constructed as to enable **it** automatically to
restore the water, and **in the** following description
it **will be** seen that this **has** been accomplished.

The next important point **is** to obtain the utmost
simplicity **of** form and **to** provide for an overflow-
passage which shall **be** both visible and accessible.
We have established **as the** second datum of our
problem that the basin be fixed and single, and have
an independent, visible, and accessible overflow-pipe.

It is important **both for** convenience and economy that the opening in the marble slab covering our basin should be **circular or** elliptic. These openings are cut **by machinery, and any form** other than **these** requires **manual work, and at once** increases the cost of manufacture. **Moreover, this form of** opening occupies the least space on **the slab** and presents the **most** agreeable effect. **A third** datum in **our** problem **is,** therefore, that **the** usual round or elliptical **form of** the opening **in the** basin-slab **be retained.**

Finally, **as the overflow-outlet must be near the top of** the **basin, some form of passageway which shall extend from** the **top to the bottom must be provided, and** since this cannot be **on the outside of** a fixed basin without being concealed **by the** slab, it must be on the inside. Hence, **as a** fourth datum, **our overflow must** have the general **form** of a stand-pipe, **and to be** completely **out** of the way of the user **it must set in a recess** under the slab at the **back of the basin,** which **must be perpendicular** at **this** point **to** receive it.

Fig. 28 represents the elevation and plan **of our** stand-pipe overflow-basin, designed in accordance with these data. **The** opening in the marble slab is circular or elliptical. A smaller circle represents the stand-pipe in a small recess **at the** rear of the bowl under the slab. The recess is large enough to allow easy cleansing without moving the stand-pipe ; and yet not so large as to injure the appearance **of** the bowl. The **bottom of the basin pitches** slightly from the centre toward the outlet ; enough to thoroughly drain off the water at each discharge, and yet not so much but that **the** last part is retarded until the siphon formed by the main body is broken by the air. The result is a restoration of the water-seal of the trap, **in case the trap be** of **a kind which**

would in basins of ordinary form be destroyed by
self-siphonage.

The diameter of the brass outlet at the bottom of
the basin measures 2 inches in the clear, in order
to allow for the obstruction caused by the
strainer. The stand-pipe is raised and lowered by a simple weighted cam, arranged as shown

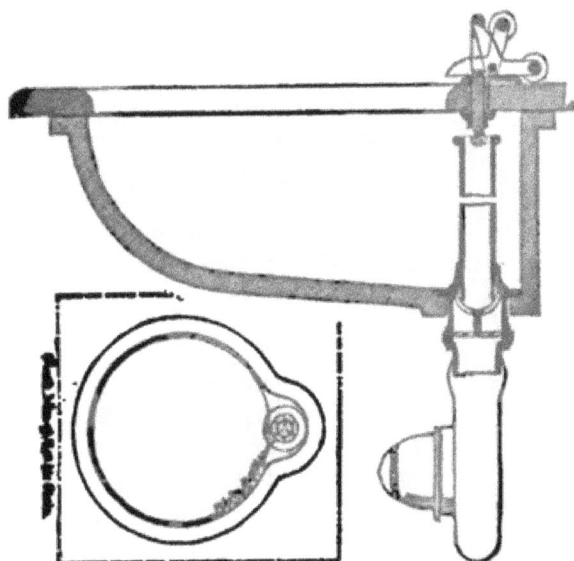

Fig. 28. The "Sanitas" stand-pipe overflow-basin.

in Fig. 29. The cam is pivoted close to the rod
which raises the stand-pipe, and its bearing surface has the form of a parabola. Its proportions
and arrangement are such as to enable it to raise
the stand-pipe without perceptible friction. The
weight of the handle, about 6 ounces, under a
leverage of only an inch and a half (the length of
the horizontal bar), is sufficient to overbalance the
stand-pipe and hold it raised. When the weight
is lifted, the stand-pipe is lowered and the outlet
closed. In this position the weight is directly over

the pivot, so that the plug and stand-pipe overbalance the weight and remain closed. Thus a single
movement of the hand
will open or close the outlet and cause it to re-
main in the position in
which it is left.

The stand-pipe rod
passes through, and is
guided by, a short tube
threaded on the lower
end and provided with a
nut by which the brass-
work is secured to the
marble slab. This rend-
ers the brasswork adjust-
able, that is, permits it to
be used with slabs of any
thickness, the slabs be-
ing perforated by a sin-
gle hole, as is usual for
the chainpost of ordinary
basins. The centre of the

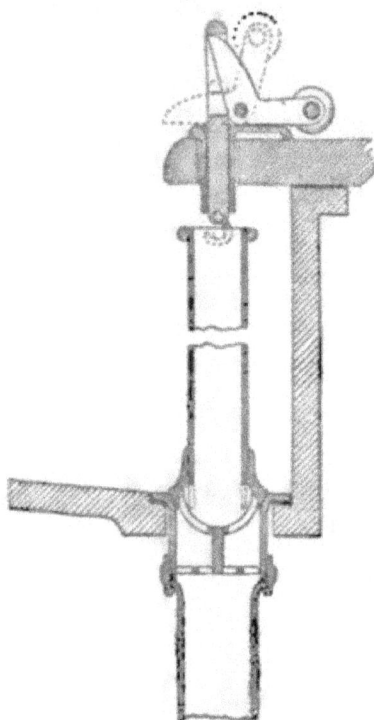

Fig. 29. Brasswork of the
"Sanitas" basin.

hole in the slab comes over
that of the basin-outlet.

In order to ensure the plug on the stand-pipe
falling always into its socket, the strainer is
attached permanently to the stand-pipe plug, as
shown in the drawing. It thus serves as a guide for
it, and allows its being lifted out at pleasure
for cleansing purposes. It is thus possible to
remove the strainer, and reach the inner surface of
the waste-pipe as far down as to the trap itself.
Thus we have an apparatus, every part of which,
both inside and outside, is visible and accessible
without unscrewing or taking down any part of the
set fixture.

The stand-pipe with its plug and strainer may be lifted out by simply unhooking it from the stand-pipe rod.

Figs. 30 and 31 show the appearance of the apparatus in perspective. By these it will be ob-

served that the recess in the basin and the stand-pipe are covered by the marble, and do not interfere with the general form seen from above. The stand-pipe measures about $3\frac{1}{2}$ inches in height and $1\frac{1}{4}$ inches in diameter. Hence its exterior surface measures $13\frac{1}{2}$ square inches. Its superficial

Fig. 30. Perspective view of "sanitas" basin.

area is, therefore, not so great as that of the ordi-

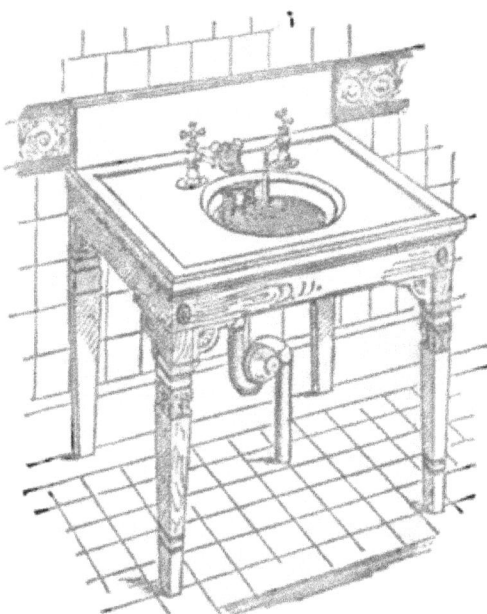

Fig. 31. Basin-set "open."

nary basin-chain. But while the chain cannot be

cleansed on account of its intricate form, the smooth surface of the stand-pipe, on the contrary, can be surrounded and polished by a single movement of a cloth or sponge. Ample room for the scrubbing-cloth is provided between the stand-pipe and the walls of its niche, so that both may be cleansed without lifting out the former.

WASHBASINS, BATHTUBS, SINKS, ETC.

The criticisms we have made on washbasins apply equally to bathtubs. No better flushing apparatus could be devised for the branch waste-

Fig. 32. "Sanitas" bath.

pipes than a properly constructed bathtub with a large outlet. As we have found the case with washbasins, so with bathtubs: the best form of discharge and overflow is the open stand-pipe; and the most convenient method of operating it is by the weighted cam already described. The cut, Fig. 32, shows the arrangement in perspective. Part of the floor in front of the bathtub is designed with hinges to lift upward and give access to the trap.

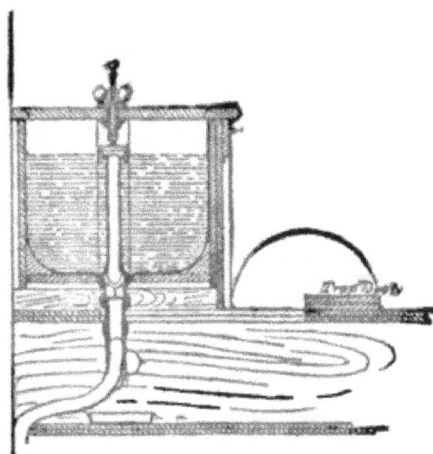

Fig. 33.   Transverse section of tub.

The stand-pipe sets in a recess at the back of the tub. The whole length of the tub is thus rendered available, and no part of the outlet mechanism stands in the way of the feet of the bather.

In ordinary bathtubs the stand-pipe is set six or eight inches from the back into the tub. It has to be lifted out by hand when the tub is emptied, and a place found for it. This constant shifting of the stand-pipe is not only troublesome but liable to cause injury to the tub itself. In careless handling the heavy pipe is often dropped into the tub, whereby the thin copper is permanently dented and marred. By the "Sanitas" fitting all handling is avoided. The supply-cocks may, of course, enter the upright end of the tub, or be placed above it out of the way as shown in the drawing.

Figs. 33 and 34 show the tub in section, and Fig. 35 in plan. The finished wooden top or slab of the tub has the usual form, as shown by the dotted lines in the latter figure. The tub itself has also the usual shape, except as to the semicircular recess for the reception of the stand-pipe.

In all plumbing work it is of the first importance that *every inch should be accessible,* and, where possible, every inch *visible* without unscrewing or undoing any part of the work. Therefore whenever a bathtub is cased in, all parts of the piping should be accessible through hinged panels.

Fig. 34. Longitudinal section of tub.

It will always be found best to follow the custom of many plumbers and builders of raising the bathtub four inches or more above the floor on pieces of studding, as shown. This renders access to the trap much easier. A bathtub raised six or eight inches would be still better. It should stand 2 feet, 2 inches, or 2 feet, 4 inches, above the floor up to the top of the finished woodwork. This would require the use of studs 6 inches or 8 inches thick under the rough woodwork of the tub. The writer has, for many years, used a tub set even higher than this, and finds it equally if not more convenient in usage.

Fig. 35.   Plan of the " Sanitas " tub.

A small piece of sheet lead, about eight inches square with its edges turned up an inch all round and soldered to form a water-tight tray, may be set under the trap, as shown, to catch the water when the cup is removed for examination of the interior.   Or a saucer or dustpan may be used for the purpose. But the best plan is to place the bathtub trap *in full sight below the plaster of the ceiling of the bathroom or china-closet below.*

It is not desirable, as some writers on sanitary plumbing urge, that the bathtub should stand open on legs without wooden casing, because the bathtub must set so near the floor that the space under it if left open would be hardly more than a crack. Dust would collect under it, and though, with proper care, it would be quite possible to remove it with the proper form of swab or broom, yet we know that places so difficult of access would be neglected by servants. The very reason which would induce us to leave, in some cases, the washstand *without* casing would lead us to case up the bathtub, that is, to avoid dirt and dust corners. In either case, whether the tub be sheathed in or not, its trap, being wholly or partly under the floor-level, would have to be approached through movable panels, so that no advantage on this score would be gained by leaving the bathtub open.

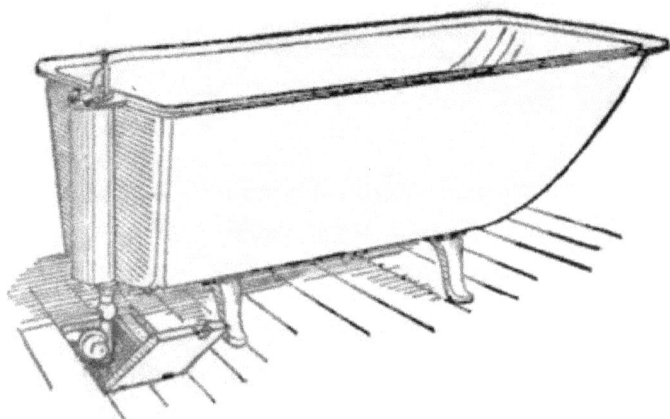

Fig. 36.   Iron bathtub with "Sanitas" waste.

In Europe where copper baths are made of metal heavy enough to stand alone without a wooden frame, the custom prevails of allowing them to stand free on the floor.   But in this country where tinned and planished copper sheets weighing only from ten to twenty-four ounces a square foot are used, a rough frame is required to support it. This requires a casing of finished wood and the casing should extend to the floor.

Porcelain baths of English manufacture are also used in this country to a certain extent and generally stand open.   Their high cost and great weight prevent their general adoption.   They have, moreover, the disadvantage of being cold to the touch and of lowering the temperature of hot water when it is used in them. Porcelain-lined iron tubs are also manufactured here and are made to stand open without casing. The liability to " scaling " of the enamel is their only serious objection. Slate and marble tubs are also made, the slabs being connected with cement.

Fig. 36 represents a porcelain-lined iron bath-tub standing open.

SHOWER-BATHS.

Fig. 37 represents a shower-bath. It stands free in the corner of the bathing-room, which has a dished or sunken floor to receive it. Jets are arranged on all sides as well as above and below.

The shower-bath is an agreeable luxury for summer use ; but for winter houses it is not so much to be recommended. The shock produced by sprays of cold water upon the body standing inactive is dangerous ; and the use of warm water in the shower-bath in winter, without intelligent precautions, is also objectionable. For these reasons shower-baths are now seldom used in city houses in the North.

The usual form of shower-bath consists in a simple rose-nozzle arranged overhead to throw perpendicular jets downwards. There are, however, a very great variety of forms and arrangements of shower-bath in use. Sometimes, in combination with the simple rose-nozzle above, a lower nozzle is provided with a rubber hose in such a manner that the lower jets may be applied in any desired direction. Sometimes the end of the bath-tub is raised high enough to enclose the shower-bath as in a niche. Lateral as well as perpendicular jets may be used within this niche, in which the bather may sit or stand. Sometimes the rose-nozzle is fur-

nished with special "needle" outlets throwing
strong sharp jets in addition to the drops from the
sprinkler. To give varying pressure the cistern is

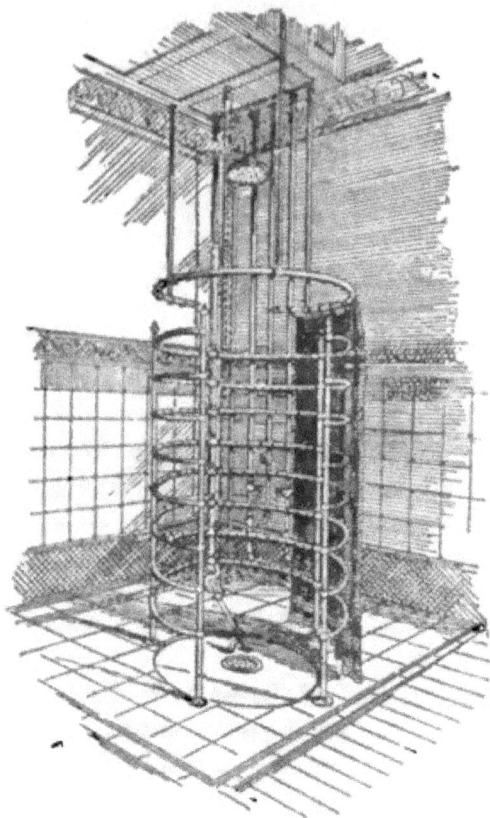

Fig. 37. Needle shower-bath.

sometimes hung in such a manner that it may be
raised or lowered at will by means of pulleys.

Fig. 38 illustrates the manner in which the
overflow-passage of a bathtub may become foul.
Colonel Waring, in his article on "The Principles
and Practice of House Drainage," in the November
and December numbers of *The Century Magazine* for

1884, speaking of concealed overflow - passages, says: "They are practically never reached by a strong flushing stream, and their walls accumulate filth and slime to a degree that would hardly be believed; . . . they are more often than any other part of the plumbing work, except the urinal, the source of the offensive drain-smell so often observed on first coming into a house from the fresh air. . . . It will, perhaps, be instructive to illustrate by a diagram the reason why the usual hidden overflow is so objectionable. . . . If we suppose the tub to be filled to the level of the overflow and its waste-plug to be removed, the

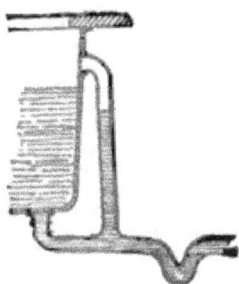

Fig. 38. Concealed overflow.

water will immediately rise in the overflow-pipe to very nearly its height in the tub. It is, of course, impregnated with the impurities of the water in the bath. Furthermore, the lighter particles of organic matter flowing through the waste will, some of them, rise by their levity into the overflow-pipe. The water rushes up into this pipe with much force, but it descends only very slowly as the level in the bath descends, so that at each operation there is a tendency to deposit adhesive matters to the walls of the pipe. What is so deposited decomposes, and escapes little by little in a gaseous form through the perforated screen into the air of the room. The amount of these decomposing matters is somewhat increased, though probably not very much, by floating particles passing through the screen when the overflowing is performing its legitimate function."

"This is the simplest statement of the proposition, and this is, perhaps, the least objectionable form of hidden overflow. **Where the** waste-pipe **is closed** at the bottom **of the overflow** by a plug **or valve** attached to a spindle **rising** through the overflow-pipe, — a very favorite **device** with some plumbers, — the difficulty **is in every way** aggravated, and the **amount** of fouled surface **is** much **increased.** The inherent defect here illustrated **attaches to** every **overflow** of this general character connected with any part of the plumbing work. In the case of a **bathtub** it may **very** easily **be** avoided, as shown **in the** next diagram." Colonel Waring then illustrates and explains the **ordinary stand-pipe** arrangement, and says : " Unfortunately such a **substitute for** the ordinary overflow **is not applicable to washbowls** as now made.[1] It may **be made available for pantry-**sinks if **the pipe can be so placed in a corner as not** to interfere with **the** proper **use of the vessel. If its** universal adoption for **bathtubs** could **be** secured, a **very widespread** source of mild nuisance would be done away with. Fortunately, it is **far cheaper than** any arrangement **for** which it is a substitute. **It is one of** its incidental uses that it enables us **to get** rid of **the dirty chain** attached to **the ordinary bath-plug."**

The size **of** the trap **for** bathtubs should never **exceed that of** the outlet **and waste pipe. A** $1\frac{1}{2}$ inch **trap is large enough for** any **bathtub or** washbasin. The proper **size for** traps and waste-pipes of different **fixtures** is a matter which **is** very little **understood.** This ignorance **has** its origin in the faulty construction of the outlets, which are **always entirely** too small. A very simple **and self-evident rule is** that *no trap except a water-closet trap should be larger or smaller than the waste-pipe at its most con-*

---

[1] This **was written** before the " Sanitas " basin **was put on the** market.

tracted point and no fixture outlet should be smaller (in its clear water-way) at any part than the trap. In no other way can the pipes and traps be properly flushed or the fixtures emptied with the desired rapidity. Hence, since we find the best size for waste-pipes for all lavatories is $1\frac{1}{4}$ inch or $1\frac{1}{2}$ inch in inside diameter, except in certain cases hereafter to be referred to, no trap should be greater or smaller than this. Moreover it is very important to bear in mind that the measure of a trap is the diameter of its inlet and outlet pipes at their smallest part. Traps are arbitrarily measured without regard to their real discharging capacity, but generally from the diameter of the inlet-pipe at its lower end, which is in most cases much larger than the smallest part of the trap. Many traps are sold as $1\frac{1}{2}$-inch traps which are contracted at some point to seven eighths of an inch or even five eighths of an inch.

The size of a trap, then, is evidently the size of its smallest part, since this part governs its capacity for discharge.

Guided by these principles we have the following rule : *The discharging capacity of the outlet of every fixture should be great enough to fill its waste-pipe "full-bore," and the size or capacity of the trap should equal that of the outlet.*

PANTRY-SINKS.

Pantry-sinks, like bathtubs, may be classified in the same manner we have done for washbasins ; and here again the open stand-pipe overflow is by far the best and most convenient form. These sinks should be constructed of tinned and planished copper weighing from sixteen to twenty ounces to the square foot. Iron and earthenware sinks are made ; but they are objectionable as exposing the dishes

and glassware to greater danger of breakage. For
this reason, also, cherry slabs are preferable to
marble.

Fig. 39 represents an oblong pantry-sink with
a stand-pipe overflow set under the slab as has been
recommended for bathtubs and washbasins. The
whole is set open to allow of easy inspection of all
parts.

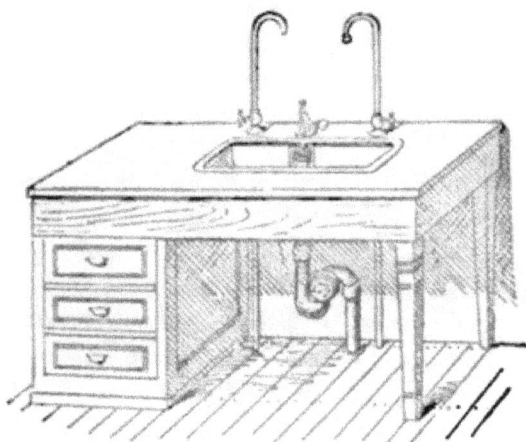

Fig. 39. "Sanitas" pantry-sink.

The waste-pipe should never exceed one and one-
quarter or one and one-half inch in diameter ; and
the outlet should have a capacity large enough to
fill this pipe full-bore. The sink should be trapped
with an antisiphon trap of the size of the waste-
pipe.

There are plumbers who still ignorantly insist
that a pot-trap is suitable for sinks on account of
the formation of grease. Let us examine the manner in which a pot-trap deals with this material.
Does it collect the grease and preserve it in its
body until such a time as it may become convenient
to remove it, or does it in some way alter its chemical

constitution so as to deprive it of its power to clog the drains? It is evident, upon the slightest reflection, that it can do neither. It is not large enough to materially cool the grease nor to retain any considerable amount in its receiver. It could not intercept as much grease, even if it were completely stuffed up with it, as would pass through the sink of a large establishment, such as a hotel, or clubhouse, in a day, or of a small house in a fortnight.

The outlet-arm leaves the body of the trap very near its top, as near as the solder jointing will allow. Hence, since grease is lighter than water, it will rise to the top of the trap and the first that collects there in any large quantity will necessarily obstruct the passage of the water. The trap can only retain in its body the small quantity of grease which would fill the corners remote from the outlet-arm, as shown in Fig. 40. The heavier matters carried into the trap along with the grease fall, on the other hand, to the bottom. These matters consist of bits of meat, bone,

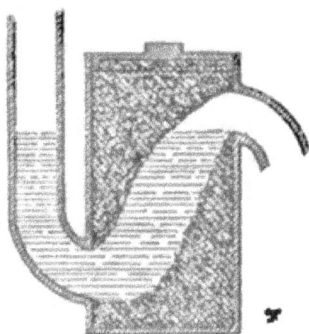

Fig. 40. Pot-trap fouled.

and vegetable. They form a foul sediment on the bottom of the trap and rapidly putrefy. The force of the water is generally sufficient to keep a passageway open for itself for a considerable length of time. Hence, after the corners have become filled with the black and rotting mass, the grease will pass through this trap exactly as it would through an ordinary S trap.

Having no proper tools to unscrew the tightly sticking cleanout cap, the owner, after vainly ham-

mering at, and disfiguring, the brasswork with such
unsuitable instrument as he can lay hands on, is
obliged to send for the plumber.

In short, it stands to reason, and is borne out in
fact, that an ordinary pot-trap has really no merit
whatever as a sink or grease-trap. It is neither
large enough to cool and retain all the grease nor
small enough to let it all pass. To cool the grease
enough to harden it before it passes into the waste-
pipes beyond the trap requires a large cesspool or
regular grease-trap. In many cases it is better to dis-
pose of the grease by sudden and powerful flushing.
The best apparatus I know of for this purpose is
the flush-pot, which is nothing more than a sinkage
with strainer in the bottom of the sink, so arranged
as to hold the waste-water until it may be be discharged, like a flush-tank, to scour the pipes and
hurry away the grease in an irresistible torrent to
the drains.[2]

Many plumbers are very fond of the pot-trap,
just as in old times they were fond of its near kin-
dred, the foul D trap, and, until lately, of the S
trap, because they were able to make them by hand
in spare moments on a "rainy day." But, in the
long run, what is best for the public is best for the
plumber. Every additional complication of the
plumbing, and everything which detracts from its
convenience, safety, and reliability, diminishes the
amount of plumbing the public will allow in their
houses, casts discredit on the art, and distrust on
the plumber.

Pot-traps, like S traps, are, however, now also
made by machinery. Moreover, new conditions
have originated new occupation for the plumbers'

[2] The "Dececo" flush-pot is manufactured by the "Drainage
Construction Co.," Newport, R. I. An excellent form of flush-pot
is described by Wm. Paul Gerhard, C.E., in his "Domestic Sanitary
Appliances."

spare moments on "rainy days." The use of the flanged iron soil-pipe furnishes him with this desideratum. The lead rings for the joints are cast in the plumber's shop as occasion demands in small, simple moulds furnished by the manufacturers for the purpose. The plumber casts in leisure moments a stock of these rings large enough to last him through the busy seasons, and finds no reason to complain of the machine-making of the pot-trap. Ultimately, it is probable that the S trap, except

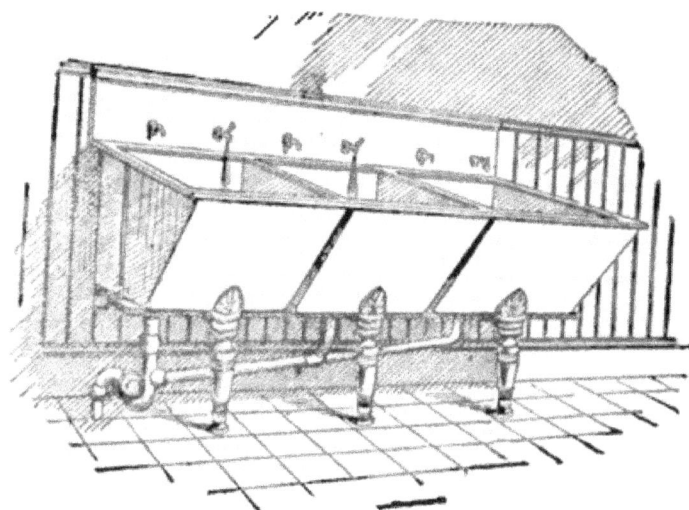

Fig. 41. Laundry-tubs.

for water-closets, and the pot-trap, machine as well as hand made, will disappear from good plumbing work altogether.

LAUNDRY-TUBS.

Laundry-tubs are made of wood, soapstone, galvanized iron, enameled iron, and porcelain.

Wooden tubs should only be used in places where their use is constant; otherwise they shrink in the intervals of disuse and become dirty and leaky.

Soapstone sinks **are** most widely used **on account** of economy and **their** general serviceableness.

The handsomest and best, **as** well as the most expensive, are the heavy porcelain trays shown in Fig. 41. A single one and a half-inch antisiphon trap is sufficient for an entire set **of** laundry trays, but no pot-trap should ever be used.

The discharging capacity or size of the bore of a trap should always be very nearly as great as that of the waste-pipes to which it is connected, in order **that its water** discharge may thoroughly scour the pipes. The fixture-outlet should also **always be** large enough to fill the waste-pipes **and** trap **full-bore**, in order to allow of their proper **flushing** every time the fixture is **used,** and also to effect **a** rapid emptying of the fixture.

When a washbasin **is** constructed with a very contracted outlet, the discharge will **be** very slow and the pipes will accumulate sediment. If a **trap** having a very small outlet, smaller even than that **of** the basin, be used with such a basin, the discharge will sometimes be more rapid than when a properly constructed trap is used, because the waste-pipe between the basin-outlet and the contracted trap **will be** filled full-bore and create **a** strong suction which will assist **in** emptying **the basin.** When the **plumber** is obliged to **set a** " Sanitas " **trap** under **one** of these ill-constructed basins with contracted outlet, and finds the discharge sluggish, the rapidity may be increased by contracting the waste just at its point of junction with the body of the trap, until he trap has **a** discharging capacity less than that of **the** basin-outlet. This practice is, however, not to be recommended. It is better to have a full-sized trap, even if the basin discharges slowly. But the **only** proper course **is to** obtain basins which have properly constructed outlets.

WATER-CLOSETS.

We have used plunger-closets in our tests for siphonage because these closets produce the se-

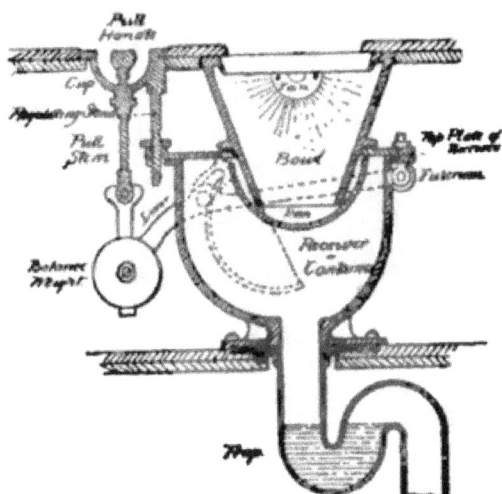

Ordinary Pan Closet
Fig 1.
Fig. 42.

verest siphoning action. It is probable that some forms of valve-closets would come very near them, however, in this effect. But valve and plunger closets, like pan-closets and mechanical seal traps, are now condemned by sanitarians as unsanitary and far inferior to simple hoppers.

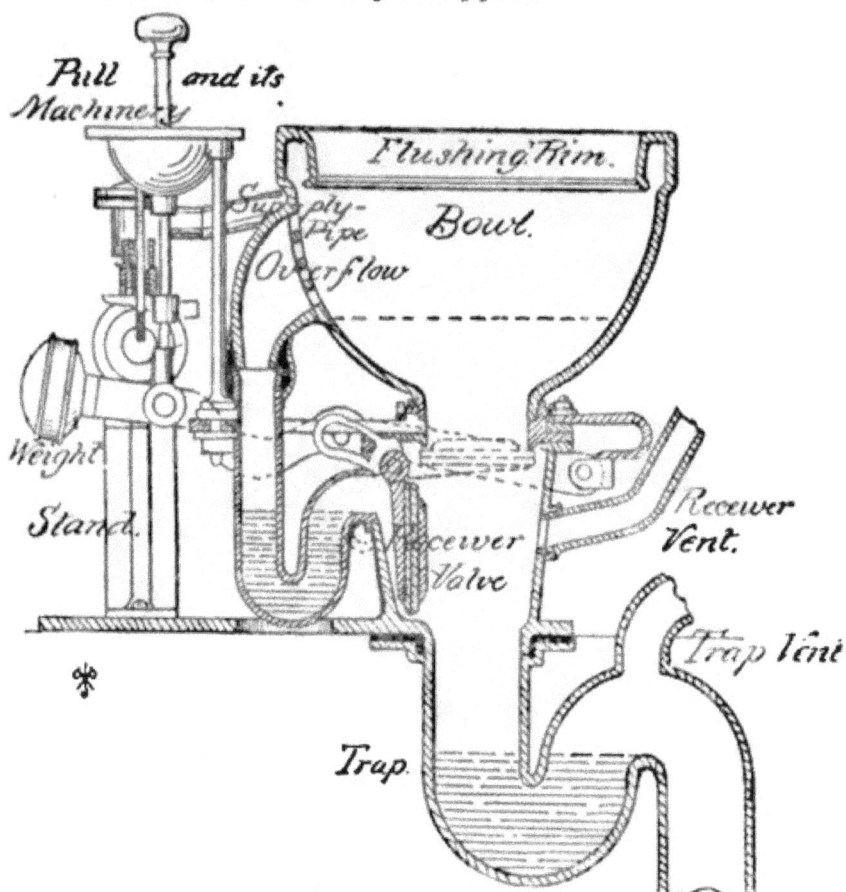

Fig. 43. Valve-closet.

Fig. 42 represents in section the ordinary pan-closet. Enough has been said by all who have any knowledge of, or interest in, sanitary plumbing about the evils of this form of fixture: its large and foul receiver, which never becomes cleansed except when the closet is taken apart and subjected to cre-

mation; its complicated form and noisy action; its
flimsy construction and its numerous vent-holes for
the admission into the house of the unwholesome
odors generated in the receiver. All are becoming
aware of the dangers arising from these defects, and
it is unnecessary for me to dwell upon them. The
one single reason why the pan-closet, with its complicated machinery and the fifty-one distinct pieces

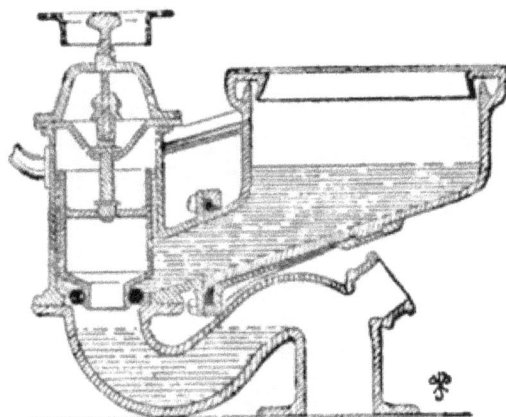

Fig. 44. Plunger-closet.

required to construct it, is sold so much cheaper
than other kinds is because these materials are of
the thinnest and flimsiest character, and no attempt
is made to prevent the diffusion of its foul gases
through the pan-journal bearings and the opening
made by the pan in usage.

Figs. 43 and 44 represent respectively types of
the valve and plunger closets. These closets are
equally complicated in construction with the pan-
closet and are equally objectionable in the theory
of their construction. Practically they are better
made, because most of those in use are patented
articles. Being of more solid and honest construction they command a higher price; but they are

very liable to leak and get out of order, and are no longer recommended by unbiased judges.

A few years ago, before systematic ventilation of the sewers and soil-pipes became universal, a tight-fitting valve or plunger might have served a good purpose in resisting back-pressure, and, as these closets were at first built without overflows, the valve or plunger performed in a measure an actual service in reducing the chances of sewer-gas leakage.

Now, however, the circumstances are altered. It is found that an overflow is necessary in these closets and this overflow-passage is rarely provided with a mechanical closure. Hence any gases which could pass an ordinary water-seal could pass through these closets by way of the overflow-passage quite regardless of and quite as easily as if the valve or plunger in the trap never existed. Moreover, the ventilation of the sewer and soil pipes renders back-pressure under ordinary conditions impossible, so that the only useful office which the valve or plunger could perform in relation to sewer-gas is no longer called for.

The valve and plunger evidently cannot prevent the loss of water-seal from siphonage, momentum, evaporation, or suction, even where the overflow-passage is closed by a ball, as is the case with some of the Jennings valve and plunger closets; for siphonage, momentum, and suction act in the direction in which the overflow-ball or valve is opened, and evaporation is chiefly due to trap ventilation. Moreover, the tightness of a valve or plunger against its seat can never be implicitly relied on. They are always liable to leak, and could never be fitted with such microscopic accuracy as to prevent the passage of any microörganisms — the bacteria, or disease-germs, or their spores— which might be in the water, through the minute openings which exist between the particles forming the valve and its seat.

The only object of the valve or plunger, therefore, is to retain a certain quantity of water in the bowl so long as they remain in working order. But it is found that this result can be accomplished equally well and much more reliably by simpler means.

The receiver or container of these closets is open to the same objections as that of the pan, differing only in degree, and the overflow-passage, not required in the latter, forms a second filth-collector, and increases the cost and complexity of the closet.

HOPPER-CLOSETS.

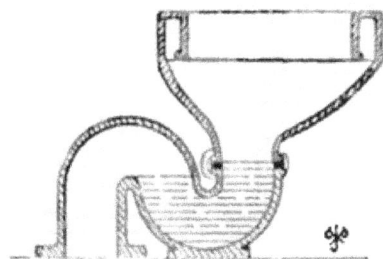

Fig. 45. The " short " hopper.

For the before - mentioned reasons sanitarians are united in condemning all mechanical seal-closets and in recommending the improved hopper - closet.

The old style of hopper, commonly designated the " long " and " short " hoppers, are objectionable as providing no sufficiently large body of standing water for the reception of the soil. The sides of the bowl in these kinds become rapidly fouled, and this form of hopper is never to be recommended except where the circumstances require the use of pails for flushing. They are not fit for the better class of houses because the trouble necessary to keep them clean will not be endured ; nor for the poorer class, because the trouble will not be taken and the closet soon becomes a nuisance in the house. Or if, by exception, cleanliness in this direction be insisted upon, the extra labor and consumption of water soon offsets the saving in first cost.

IMPROVED HOPPER-CLOSETS.

There are several forms of improved hopper water-closets, among which the best are the following : —

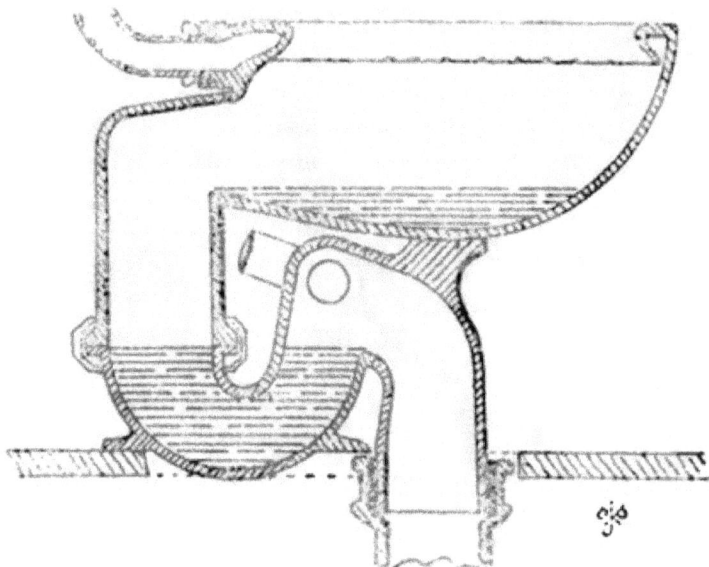

Fig. 46. Ordinary "wash-out" closet.

Fig. 46 represents an ordinary wash-out closet. It contains the large surface of standing water for the reception and deodorization of the soil. The flushing-stream sweeps across the bottom of the bowl with great force and drives the wastes before it into the trap. Whether or not the trap itself be emptied depends upon the length of time the flushing is continued after the bowl is cleared.

The objections to this form of hopper are: (*a*) the presence of the extended pipe surface between the bowl and the trap, and the inaccessibility and invisibility of the latter; (*b*) its extravagant consumption of water, the waste matters often whirling about some time in the bowl before they are driven out; (*c*) its excessive noisiness in action; and finally (*d*), the spattering occasioned by the violence of the flushing.

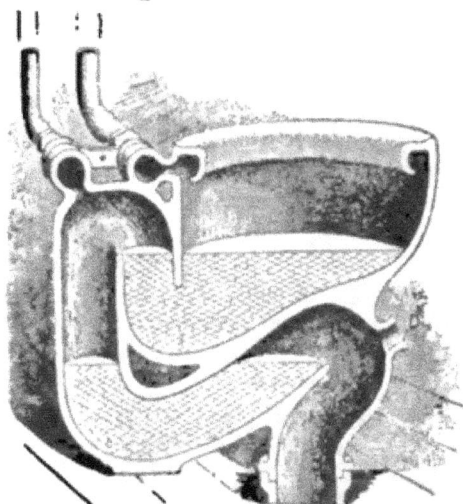

Fig. 47 represents a closet constructed with a double trap, one below the other. This water-closet, which is called the " Tidal - Wave," works on the principle of the siphon. The air between the traps is exhausted by the action of the valve and cistern.

Fig. 47.   Boyle's " tidal-wave " water-closet.

This unites the two bodies of water in the traps and forms the siphon which empties the bowl. This apparatus has lately been considerably improved by its manufacturers, and as now made forms a very excellent closet.

Fig. 48 represents the " Dececo " closet. This is a simple and effective apparatus, and works on the principle of the Field's flush-tank.

A weir-chamber is used below the trap to assist in charging the siphon. The weir-chamber is just below the floor. In order to charge the siphon the water is let into the basin through the supply-pipe and the flushing-rim until it overflows the outlet of

the trap and falls into the weir-chamber below.
This closes the inlet of the weir-chamber before it
can escape through the outlet and prevents the air
from entering the siphon. The air already there is
carried out by the current of water and the siphon
is formed. As soon as the water in the bowl
descends to the bottom of the dip of the trap air
follows it and breaks the siphon.

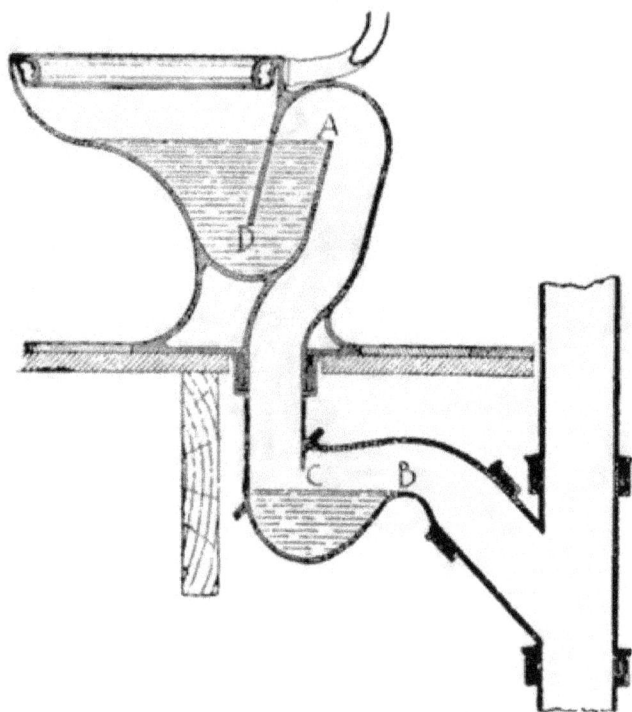

Fig. 48. " Dececo " water-closet.

The bowl is then refilled by the afterwash. This
closet is an ingenious one; it is simple and durable,
and the later and better forms seem to produce in-
variably the siphonic action in the manner described,
giving the requisite flushing without spattering or
waste of water. It should not be used as an ash-
barrel or receptacle for all kinds of rubbish. When

properly used it is a closet which never needs
repair.

Compared with the " wash-out " closet of Fig. 46,
these points of advantage are to be noted here:
(1) The depth of water in the bowl is much
greater where depth is needed to receive and deodorize the soil. (2) The trap is in sight and the
walls of the outlet are under water and are odorless
instead of the reverse, as in the " wash-out " closet.
(3) The water-seal in the trap is twice as deep
and therefore better able to resist the influence of
siphonage, etc.

THE " SANITAS " WATER-CLOSET.

Figs. 49, 50, and 51 represent in section and plan
the " Sanitas " water-closet. The form is absolutely

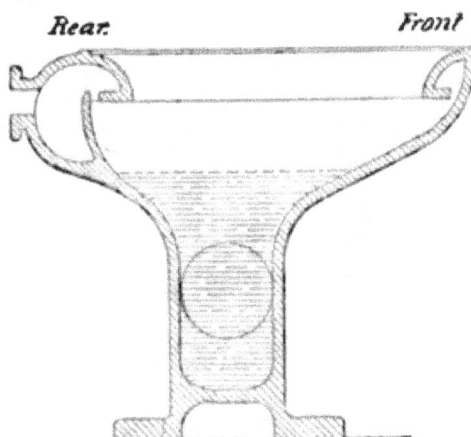

Fig. 49.   Longitudinal section.

simple. The bowl and trap are one and the same
thing, inasmuch as each forms the other. The
flushing is accomplished without machinery of any
kind in the closet, but by the pressure of the water
only, and the quantity of water required is reduced
to a minimum. The supply-pipe (see Fig. 50) enters
the bowl below the normal level of the standing
water therein, and stands permanently full of water

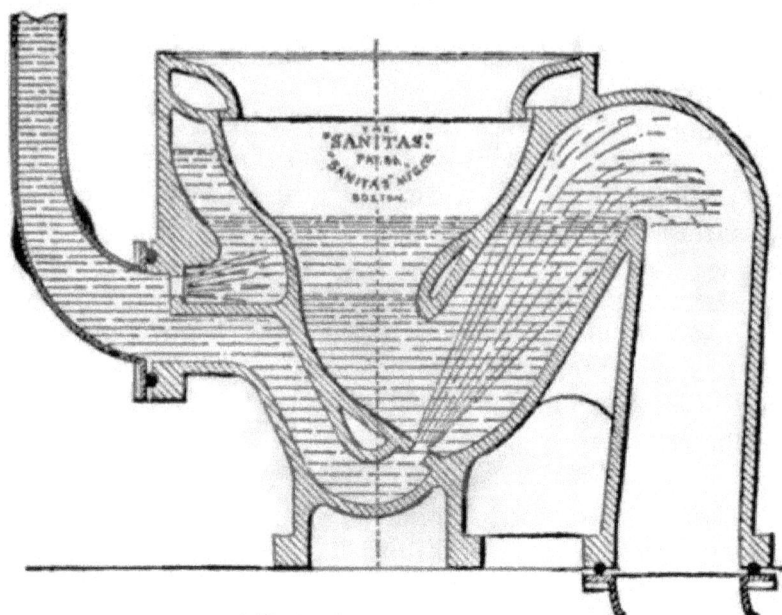

Fig. 50. Transverse section.

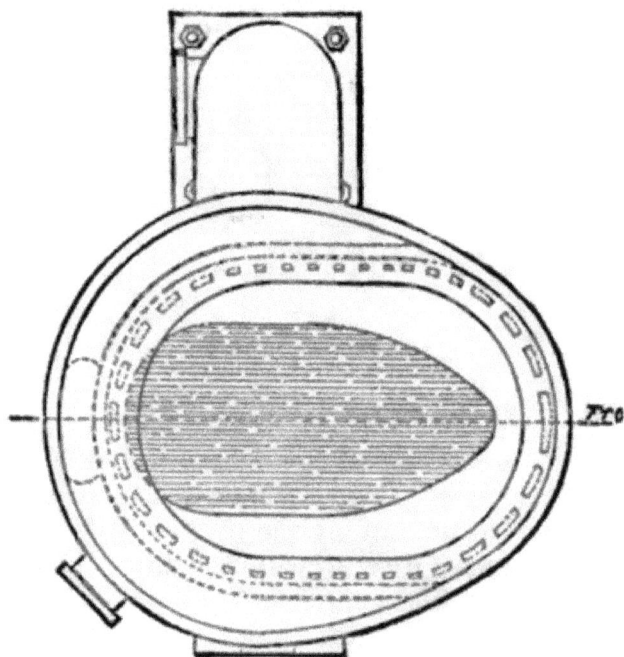

Fig. 51. Plan.

up to the cistern-valve. This water is held in the supply-pipe by atmospheric pressure, the pipe being closed at the top by the cistern-valve and at the bottom by the water in the closet-bowl. The lower end of the supply-pipe is perforated at two places

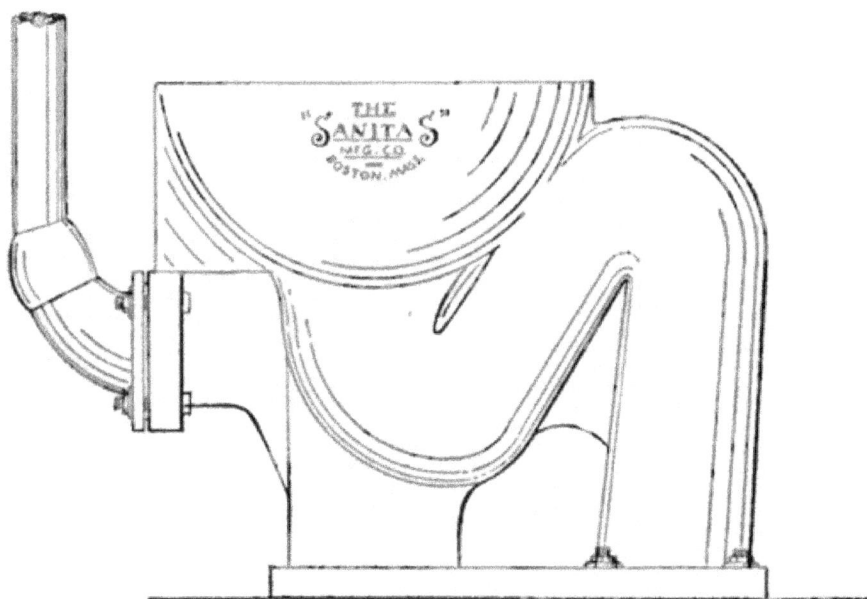

Fig. 52. Front view.

independent of each other : first, at a point intermediate between the overflow of the trap and its dip ; and, second, at the bottom of the trap. The first supplies water to the flushing-rim, and the second furnishes a jet which lifts part of the water out of the trap and bowl by its propelling power. Since both jets enter below the level of a large body of standing water in the bowl, they act noiselessly, and, as the supply-pipe stands always full, they act instantly, and the flushing of the closet is very rapid. The operation is as follows: Upon opening the cistern-valve the water in the supply-pipe is instantly set in motion by the pressure of the atmosphere on the surface of the cistern and escapes through the two orifices in powerful jets. The lower jet removes part of the water from the trap and causes the water and waste matters in the water-closet to sink into the neck of the bowl. Meanwhile the upper jet fills the passage leading to the flushing-rim, and, descending into the neck of the bowl, falls upon and drives out the waste matters collected in the neck without noise or waste of water. The action is almost instantaneous. The cistern-valve being again closed, movement in the supply-pipe immediately ceases, and the water in the flushing-rim and passages leading thereto falls back into the closet and restores the normal level of the standing water in the bowl and trap.

The form of the closet-bowl is such that the sur-face of the standing water therein is very large. It has the shape best calculated to receive and deodorize the waste matters falling into it. The water is deepest at the back of the closet, and very deep at the point where the wastes strike. All parts of the trap and bowl are easily accessible from the bowl itself, and there is no superfluous space and no surface which is not thoroughly scoured by the flushing-streams in the normal usage of the closet. There is no invisible trap below the bowl, and when the closet appears to be flushed clean it is so.

This closet can be easily flushed in one second by less than a gallon and a half of water. There are several advantages in having the supply-pipe of a closet enter below the level of the water in the bowl and closed above with a valve without air-pipe so that it shall remain always full of water. In the first place its action is instantaneous and noiseless. The water does not have to fall from the cistern to the closet before it begins to work. In the second place the friction of air in the pipe is avoided and the water exerts at once its full power in discharging the waste matters. Hence a very considerable economy of water is the result. As already stated, the upper orifice is placed below the level of the standing water in the closet-bowl, but above the dip of the trap. This position of the upper jet gives us another very important advantage. Should the water in the closet be lowered by evaporation or siphonage below the upper orifice, air will at once enter the supply-pipe through this orifice and water will then descend from the pipe into the closet through the lower orifice, until the upper orifice is again covered, and the seal of the trap is thus automatically maintained by the water in the supply-pipe. This pipe may be made capacious enough to restore the seal as often as it is likely ever to require it. A pipe $1\frac{1}{2}$ or $1\frac{3}{4}$ inches in diameter and six feet long will contain water enough to secure the seal against destruction by evaporation for a great many months, even in the dryest and hottest weather. Hence the closet may be left to itself in city houses for the entire summer's vacation, without fear on this score, and the danger of a loss of seal through siphonage is also reduced to a minimum. The seal of this closet is over three inches deep. Such a seal is difficult to break by siphonage, even without the use of our automatic supply-pipe, which I have called the " Sanitas " water-closet supply-pipe.

It will be observed that the closet is provided with a ventilation opening near the crown of the trap. This ventilation will seldom if ever be required to prevent external siphoning action. But it is useful to break the siphon which would be formed in the closet-trap itself during the flushing, and thereby prevent the noise which the formation of such a siphon would occasion in use.

Hence, where it is desired to have the closet act noiselessly, a short, two-inch vent-pipe should be used to connect the crown of the trap with the soil-pipe immediately on a line with it or with any other convenient point of the soil-pipe near by. Moreover, the law at present in some cities requires every trap to be vented regardless of consequences. So that such a vent-opening may at times be needed to conform to the requirements of this arbitrary and ill-considered provision. It will also be observed that the closet is provided with a cistern-overflow connection which may serve also when desired for a bowl ventilation pipe connection.

An important advantage in having the trap and bowl of a water-closet combined in this simple form is that they may be easily emptied in winter to prevent freezing. This is particularly desirable in the case of summer residences which are closed up in winter. The water may be easily sponged or pumped out of this closet without taking it apart, whereas closets having inaccessible traps under the bowl or floor cannot be emptied or cleansed without taking the apparatus to pieces, and in the case of many forms of wash-out closets where the trap under the bowl is in a single piece of earthenware with the bowl, the emptying or cleansing of the trap is either very difficult or altogether impossible.

The upper flushing is accomplished without spattering because the pressure of the upper jet is relieved at the upper orifice, and the water quietly overflows the rim of the bowl. Fig. 52 gives a front view of the " Sanitas " closet.

LATRINES AND TROUGH WATER-CLOSETS

are designed for use in public places where an
attendant can be employed to take constant charge
of them, and where water is so abundant that its
extravagant consumption is no disadvantage.
Trough water-closets consist of a long reservoir or
trough, inclined toward one end, where a discharge-
plug is placed, and having a single or double row
of water-closet seats placed over it, so that all the

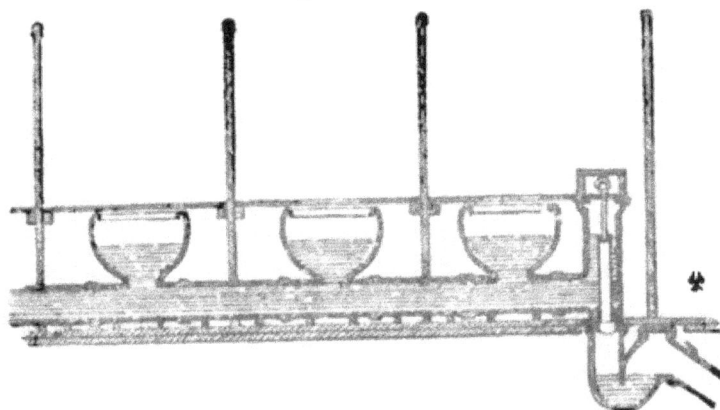

Fig. 53.   Latrines.

closets are flushed together, or, in other words, so
that the flushing of one necessitates the flushing of
all the rest in the series connected with it.   They
are constructed in different manners, either of
brickwork having vertical sides and rounded bottom,
or of iron, usually enameled.

. Latrines (Fig. 53) are practically trough water-
closets having the trough diminished in size, and a
bowl or funnel discharging into it under each seat.
The bowls are constructed of earthenware or white
enameled iron, and the trough or pipe with which
they are connected is made of iron, and has a trap
at its end under the discharge-plug.   In the figure

the discharge-plug is **hollow, and consists of a** stand-pipe with overflow-**passage** through it. The height of **the** overflow **regulates** the position of the standing **water in** the **bowls.** The plunger or dis-charge-plug **is** under **the control of the** attendant, who flushes **the closets as** often **as** he considers it advisable. The bowls **are so constructed** that the waste matters fall directly **into** the standing water, and nothing strikes their **dry** sides; they are thus **partially** deodorized. But **the** liquid and soluble **portions of the solid wastes, which are** allowed by **the faithful attendant to remain for some length of time in the latrines, as well for the sake of** economizing **water as to enable** him to attend to his **other** duties, **soon** precipitate a slimy **deposit all along the** inner surface of the closet, and particu-larly around the plunger-chamber. This is not easily removed, and always forms more or less of **a** nuisance. In most cases it will be found much better to provide, instead of latrines, a row of good hopper-closets, **with** treadle, **door, or seat attach-ment for** automatic flushing, **if desired.**

SLOP-SINKS, SLOP-HOPPER **SINKS, AND SLOP-HOPPERS.**

Figures 54 **and 55 represent two kinds of fix-tures** designed for **the** reception of slops. **These have no means provided for** the flushing of the **walls** of **the sink.** Either **may be** provided **with a** flushing-**rim for** the **purpose.** But **the use of the** flushing-rim in private **houses is** oftener neglected than observed. Servants will **not take the** trouble **to** thoroughly cleanse the slop-hopper **at** every usage, and it **soon** begins to emit **a disgusting** odor. In hotels or **large clubhouses, where their use** is constant **and under systematic** supervision, where special **attendants are detailed to** take charge

of them, and where each story is independently provided with a separate slop-hopper, their use may
be recommended; but in private houses they should

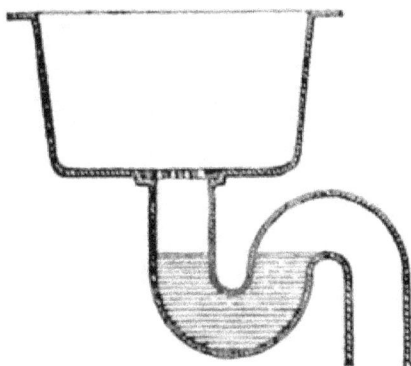

Fig. 54. Slop-sink.

never be allowed. A good hopper water-closet,
with a strong enameled iron drip-tray to protect

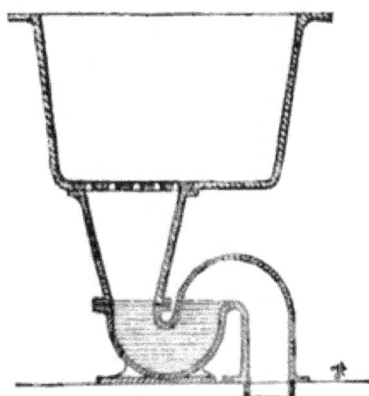

Fig. 55. Slop-hopper sink.

the bowl, is much better, inasmuch as, while it
serves the purposes of the slop-hopper equally well,
it escapes its objections in ensuring a periodic flush-

ing. Every time **the** closet is used for the purposes of nature **it is** thoroughly flushed, and even
slops are much seldomer allowed to stand in the
bowl, because their presence **would be** immediately
detected by **the** next regular **user** of the water-
closet, and **the** damage would **be** likely to " recoil
upon the head of the offender." It is customary **in**
private houses to place the slop-sink in **the** attic,
but no house-owner can give any better **reason** for
its existence **than that** he had seen it in some other
houses. When valve, pan, and plunger closets **were**
used to the exclusion of the more modern **hopper,**
the slop-sink had a certain *raison* *d'être.* In these
closets, especially those requiring **an overflow-pas-**
sage, the closure **of the outlet is apt to cause an**
overflow of **the** slops **when** a large pailful **is** poured
in quickly. But the modern hopper-closet has a
clear, open passageway into **the** drains, and, being
provided with **the** most improved form of flushing
apparatus, is, in fact, **the best form** of slop-hopper
that has **been** devised. **Some** persons **who have**
insisted, even contrary **to the** advice of **their archi-**
tect or sanitary engineer **(who** now **unite in** condemning them), upon **having** the customary **slop-**
sink duly installed in their attics, wishing **to have**
at least an appearance of a reason **for** their way-
wardness, urge **that the** virtue of **the slop-sink** lies
in the strainer : this serves to prevent **the obstruc-**
tion of the drain by scrubbing-brushes, rags, large
cakes **of** soap, or other household articles **used** in
scrubbing, capable of clogging **the** soil-pipe, which
a careless servant might throw with the slops into the
sink. This office of the strainer **is** certainly **a** useful one, and if every story in the house contained a
slop-sink provided **with such a** guard, and every
water-closet had a **movable or** portable strainer

endowed with sufficient intelligence to close the outlet only when slops were poured in, the soil-pipes
might really be protected from **the gross** careless-
ness our friends so much feared. But **as such** a
profusion of slop-sinks and **strainers is** evidently
impossible in private houses, **and as** slops are collected in every story of **the house as** well as in **the**
attic, and **as no servant careless** enough to throw
scrubbing-brushes **into a water-trap** would take **the**
trouble **to lug slops from** the lower stories up to **the
attic, in order to** protect the neighboring water-
closet trap from such an **accident,** or, in other
words, mount one or more flights **of** stairs **to** avoid
the trouble **of** removing the **scrubbing-**brush from
the slop-pail before emptying **the** slops into the
nearest water-closet **bowl, it is** evident that the
argument of **protection to soil-**pipes **has** little
weight.

For hotels and some other **public** buildings, **the**
slop-sink should have a good flushing-rim. It then
becomes the so-called " slop-hopper," and the bowl
should **be** properly protected **by a** stout iron drip-
tray, properly supported, **to receive the** frequent
blows bestowed **upon the hopper by the careless**
pail.

URINALS.

As they are generally made, urinals **are very objec-**
tionable things **in private** houses. Urine undergoes
rapid decomposition, and gives off a powerful and
very disgusting odor. When in this state, it has
the power **of** turning fresh urine into the same condition almost immediately, so that **unless** the urinal
is so formed **and** placed that its surfaces are thoroughly cleansed after use, it soon becomes a very
foul and disagreeable fixture in a house. Fig. 56

represents the most economical form of urinal as they
are now made. The bowl is generally constructed
of glazed earthenware, with
some form of fan or flushing-
rim for spreading the flushing stream over its entire
interior surface. The urine
escapes through numerous
perforations in the bottom
and back of the bowl, into
the waste-pipe. In some
forms the trap is made in a
single piece of earthenware
with the bowl. There are
a number of different forms
of urinals, both swinging
and stationary, and they are
flushed either by a stop-cock
directly on the supply-pipe,
to be turned by hand, or by
a special cistern. The former method of flushing is
open to the same objection
as the direct supply to water-
closets, and is now forbidden
in some places by law. The pressure may be at
times insufficient to fill the pipes, and the foul
air from the surfaces of the urinal, perhaps containing disease-germs, may be sucked into the supply-
pipes on opening the stop-cock. In the figure, an
automatic flushing cistern is used, which has within
it a tilting vessel arranged to give a periodic flush
as it slowly fills under a small faucet kept constantly
open. This is perhaps the only certain method of
ensuring a sufficient flush for single urinals constructed in the usual way ; but it involves a great

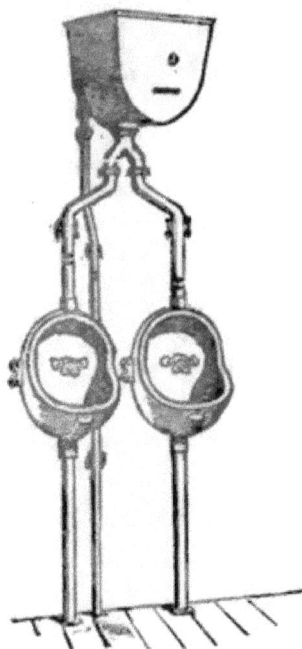

Fig. 56. Urinals.

consumption of water and is very wasteful, inasmuch as the flushing goes on always, whether it be
required by the use of the urinal or not.

For private houses it is much better to construct
the urinal in the manner shown in Fig. 57. It is
a simple hopper-closet raised to the height of a
urinal. By this arrangement, all of the advantages
of a urinal are obtained, without any of the objec-

Fig. 57. Combined urinal and slop-sink.

tions. Moreover, by stepping on the steps or foot-
rests at the floor in front of the fixture, the device
serves equally well as a water-closet. The writer has
found by experience that this form of urinal never
becomes foul, nor is its use as a water-closet accom-
panied by the least inconvenience. He has used it
both in public and private buildings with equal

success. The bowl, containing a large body of standing water, dilutes the urine, and prevents it fouling
the sides. Habit, with water-closets, leads to its
flushing after its use as a urinal at times when the
ordinary form of urinal would have been left un-
flushed. But should, by any chance, the flushing be
neglected, the next use of the fixture as a water-
closet would insure its cleansing. Moreover, by
combining the two fixtures in one, economy both of
space and first cost is obtained, while the offensive
appearance and smell of the urinal is avoided and
the consumption of water is greatly diminished. Not
the least of the advantages of this arrangement is
that it is suitable for use by both sexes, a consideration of some importance, especially in the hall of
a private house, where the want of space limits one
to the use of a single fixture.

In public buildings, however, such as hotels, rail-
way-stations, manufactories, school or club houses,
where proper and systematic attention may be ex-
pected to be given to them, urinals may become not
only desirable, but absolutely necessary. Stall
urinals should also be constructed in various places
in the main thoroughfares, easily accessible to the
public, as an important sanitary measure.

PART III.

SOIL AND DRAIN PIPES.

The material for our pipes naturally forms the first subject for consideration, inasmuch as upon it their proportion, treatment, and arrangement in a great measure depend.

By far the most suitable material yet discovered for soil and house drain-pipes is iron, and the most important matter connected with its use is the formation of the joints between the separate pieces.

Lead has been almost entirely abandoned in this country for soil and drain pipes, on account of its want of strength and rigidity, its comparative high cost, its liability to be perforated by vermin, nails, or corrosion, and of the greater time and labor required to make the joints. Large lead pipes often sag of their own weight and tear away at their points of support. The action of alternating hot and cold water also produces a destructive effect upon the material. In England lead soil-pipes are still used, but it is not customary to use the soil-pipe for the conveyance of all kinds of waste, and hot water from lavatories and sinks is carried into separate pipes, so that the material as used abroad is less objectionable.

S. Stevens Hellyer, the well-known and popular English writer on sanitary plumbing, says, in speaking of the question as to the material most suitable

for soil-pipes : " This may seem a curious question
to ask of plumbers — as well ask a shoemaker of
what material should boots and shoes be made!
Everybody knows that the latter would say:
'There's nothing like leather,' as the former is sure
to say : 'There's nothing like lead.' . . . Allowing
experience to be my schoolmaster, I answer lead,
especially for our climate." Mr. Hellyer claims the
following points of superiority for lead : its greater
smoothness, greater resistance to corrosion, greater
ductility for bending to suit the various positions it
has to occupy, more perfect jointing, greater adaptability for connecting with branch wastes, and
greater compactness, which allows it to be placed in
slots or niches smaller than those which are required
for iron. He admits the following objections : its
deterioration under alterations of temperature which
tend to work it until it breaks, its sagging, its ex-
pensiveness, its liability to be perforated by rats or
carpenters' nails, its greater weight, and the requirement of greater skill in making the joints.

The advantages which Mr. Hellyer claims for lead
have within late years lost their force. Improved
methods of protecting and jointing other materials
have placed them in these respects far ahead of
lead, as will be shown hereafter. White enamel is
now applied as an inner coat to cast iron in such a
manner as to render the inner surface as *smooth* as
that of new lead. But in use lead soon loses its
smoothness, the sewage adheres to the surfaces of
the pipe to a greater or less extent and roughens it,
in time, with a hard deposit of greater or less thickness according to the usage of the pipe, so that the
difference in smoothness at the outset in favor of
lead as compared with an iron pipe, properly coated,
is of small consequence after a few years' use.

The **numerous cast** bends and fittings now **made and** adapted **to** every possible turn **or** angle **liable to be** encountered in arranging **the pipe** renders the *ductility* of the **lead** pipe **no** longer **of** any advantage. Finally, other and more suitable materials are now jointed in such a manner **as** to render them quite **as** *compact* as the lead **pipe.**

Stone and ***brick*** **drains cannot** be effectively flushed **on account of the roughness of their** interior surfaces. **Moreover,** they are porous to a certain extent, **and** the cement with which they are **laid** is always **more or less** pervious to water.

Wooden drains soon decompose and leak, and when made of **plank** must be of such a section that scouring is impossible.

Copper is easily **corroded by the acids of sewage and** decomposition, **and it is, moreover,** too **expensive when made** heavy enough **for the purpose.**

Zinc, tin, **and** *galvanized iron* are **totally unsuitable**, and **not** to be considered for a moment. In **the** worst kind **of** so-called "Gerry buildings" they **are,** however, occasionally **used.**

CAST IRON.

Cast iron is the material **which in** this country **for** the last twenty years has been most generally used. It has in this time proved itself to be a most reliable and excellent material for soil-pipes. It is light, cheap, stiff, **and** strong, and **it** corrodes so slowly that, **if** of the proper thickness and quality **of** iron, and properly cast, coated, **and put** together, **it** will last as long as the house. **The** inconveniences at present attending its **use** are not inherent **in the** nature of the material. As now **made, the** pipes are often cast of uneven thickness, and they are always improperly jointed. Neither defect is **necessary.**

The experiments of **M.** Gaudin, **made** in 1851, show **the** maximum **rate of** loss by **rust of** uncoated cast-iron pipe exposed to the action of clean, fresh water on both sides to be a **little over** an eighth of an **inch** a century. His experiments extended over a period of thirteen years. **With the** present methods **of** protecting iron, its life **can** be very greatly prolonged; indeed, even the use **of** the ordinary bituminous coating (coal-tar pitch) has proved, **when** it is properly applied, to be able to keep **the** pipe quite intact for twenty years. The life of a soil-pipe, **even** when quite thin **and uncoated,** has been found **by experience to be so great that it is not unreasonable to suppose that the greasy** matters **contained in sewage serve to** protect the **pipe** in a measure from **the** water and from **the** corrosive **action of the acid** components **of** the sewage.

JOINTS.

Equally important with **the** question **of the** mate-rial is **the** manner in which the several parts are **put** together, inasmuch as upon this depends not only the safety **of** the work, but also, **in a** measure, the choice of the material itself. **The** question of *joining* **or** *coupling* the pipe will therefore next be considered.

The ordinary joint is neither tight nor permanent; **it cannot be** made **to** resist water **or** gas under pres-**sure, and it is** soon **destroyed by** alternations of **heat and cold in** the pipes, such as are often produced **by the passage** through **them of** hot water **or steam. It is expensive** both in **time** and material. **It re-quires expert** labor **to** adjust, **but defies expert labor to take it apart** again without **more or less** destruction **of** the **piping.** Even the **process of**

putting together involves a hammering which endangers the integrity of the pipe, and the most experienced and careful workman often cracks it in the process. The safe use of white enameled pipe is out of the question with the calked joint, because the jarring produced by the calking-tool cracks the enamel.

Fig. 58 shows the ordinary hand-calked joint. It is made with lead and oakum or jute. A gasket of jute or other similar fibre is inserted into the cavity of the bell or hub, and the spigot end of the length next above it is set firmly down upon it, or the gasket is rammed in with a tool after the lengths are set up. The gasket is used to prevent the lead from running out of the joint and obstructing the bore of the pipe at some point below, besides wasting the lead. The lead is now poured upon the gasket from a ladle and shrinks as it cools. The calking-tool must then be used to expand it again and drive it into the cavities and pores of the iron. A faithful and skilful operator can by perseverance succeed in fitting the lead into the iron at all parts of its circumference, so as to make it tight for a time, just as a painstaking dentist can drive the gold by patient labor into the cavities of a tooth, and temporarily arrest its decay. But the process in both cases is slow and uncertain. The dentist confines his calking to a single small spot well within his reach, and he labors with extraordinary care. Yet the filling often fails when put to the test. The plumber must work quickly over an extended field often in awkward positions; he must perform a delicate task with clumsy tools.

Fig. 58. The ordinary bell-and-spigot joint.

The metals to be welded together are often so placed that it is impossible, without the utmost patience and skill, to reach them properly. The result is that when put to the test the joint *almost always* fails. Extra heavy pipe and hubs are required to withstand the blows of the calking-tool. Lighter pipes cannot be made tight without danger of cracking the iron. It is now generally recognized and acknowledged that a plumber's calked joint is rarely either air or water tight, though a vast amount of lead and labor is spent on them to make them so. When we reflect that the sole aim and object of a soil-pipe joint is to make a gas and water tight connection between pipes, we see that the method commonly employed is an absurdity, and reflects little credit upon human ingenuity.

Even supposing that, by chance, a calked joint has been made to stand the test which is now properly required of it when new, its tightness is very soon destroyed by the expansion and contraction of the pipes caused by the passage through them of hot water or steam. The expansion of the spigot is in such cases greater than that of the hub, because it is on the inside nearer the heat and not protected like the latter from the hot fluids passing through the pipes. Hence the lead is temporarily compressed between the spigot and hub, and, being inelastic, does not resume its original bulk when the pipes cool again. A minute opening is thus formed all round the spigot, as shown in the lower branch of the pipe in the cut, and the joint leaks.

The object of requiring the whole system of pipes used in plumbing a house to be filled with water as a test is not only to determine the tightness of the joints in a manner which is impossible with the peppermint or smoke tests, since these can be eluded by a temporary coat of paint or putty, but also to try the quality and thickness of the metal. If a pipe is very defective in casting, its weakness will be revealed by a strong pressure test, and the faulty piece rejected.

Another serious objection to this joint is the difficulty of disjointing pipes in which it is used. The usual way to take out a pipe, once so put together, is to break it to pieces, and then remove it by degrees. There is, in fact, no practicable alternative; for to melt off the lead would not only be expensive and dangerous, but involve the disjointing of quite a considerable length of pipe in order to enable a single spigot to be lifted two inches, or enough to disengage it from its hub. Now alterations in our plumbing arrangements are necessarily so frequent that this objection becomes a serious one.

The necessity of using fire in a house in process of construction for melting the lead necessary to make this joint is also a formidable objection to it, on account of the danger of igniting the surrounding carpenters' litter and burning down the house. It is true that lead or solder melting would have to be carried on for other purposes, such as wiping the joints on the smaller pipes, but the less use we have for the solder-pot the less will be the danger, and the less the temptation for the workman to carry on the melting in dangerous places in order to save himself the trouble of running up and down wearisome flights of stairs to a place of safety.

Still another very serious objection is the temptation this joint opens for *fraud*. The lead may be partially or even wholly omitted without very great risk of detection, since it is out of sight, and frequently immediately covered by a coat of paint.

The calking may be still more easily slighted. If the hydraulic test is not demanded, a coat of paint or a little putty will easily make the joint stand the smoke or peppermint test. A few of the joints well within the reach of the house-owner may be filled with genuine lead, while all those which are covered by floor boards, or are not easily accessible, may be composed of paper and sand, and covered with putty. Possibly a thin coating of lead may be poured on top to present an honest appearance, and satisfy the suspicious and shrewd house-owner who goes about probing the nearest joints with his pen-knife in order to ensnare " the rascally plumber."

Finally the bell-and-spigot joint, when faithfully made, is very expensive both in material and labor. The amount of lead required for each joint, including waste, is estimated at about a pound for every inch in the diameter of the pipe. Thus an ordinary four-inch soil-pipe consumes four pounds of lead in each joint.

The average length of time required by a skilful pipe-layer to make a single joint is estimated at twenty minutes, not including, of course, the planning of the pipe system or the cutting and general arrangement of the pipe sections for their proper positions, a part of the work which has no connection with the kind of joint used.

THE " SANITAS " PIPE JOINT.

Our " Sanitas " joint has been designed to obviate these defects and enable lengths of cast-iron pipe to be securely and economically united.

In general terms it may be described as an *adjustable flanged joint* with lead washers or gaskets for packing. It is a steam-fitter's joint, with improvements which adapt it for use in plumbing.

The leaden gaskets are star-shaped in cross-section, and are crushed between the flanges of the

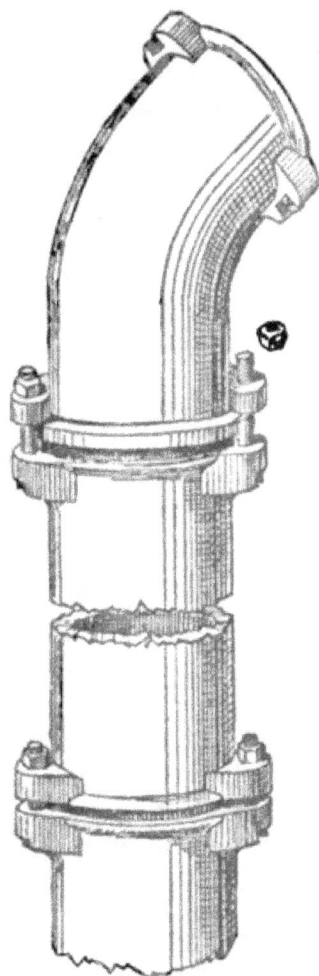

Fig. 59. The "Sanitas" pipe.

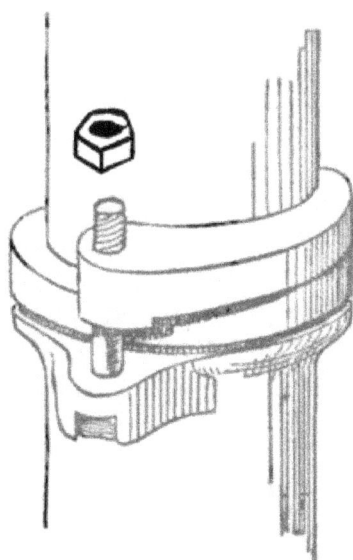

Fig. 60. Detail of joint.

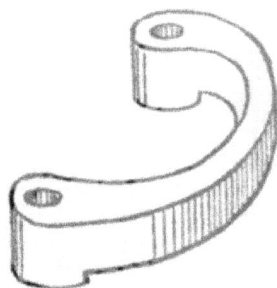

Fig. 61. The half-ring.

pipes to be connected by means of bolts and nuts.

Figs. 59 and 60 illustrate the joint. In Fig. 59 the lower form of the joint is to be used in connecting straight pipes, and the upper form in con-

necting branches and bends with each other and with
straight pipes. The latter differs from the former
in the addition of a half-ring just above the flange
of the bend or branch, and is illustrated on a larger
scale in Fig. 60. This half-ring is shown in detail
in Fig. 61, and permits the bent or branched pipe to
be revolved about its axis in setting, before the
bolts are tightened up. The entire pressure is
brought upon the ends by the small shoulders on
the half-rings in such a manner as to prevent their
fracture when the bolts are tightened up. The bolt-
heads set in square recesses in the flange ears to
prevent them from turning when the nuts are
screwed home.

Fig. 62 shows the lead packing-ring in perspec-

Fig. 62.   Lead packing-ring.

Fig. 63. Sec-
tion of packing-
ring.

tive, and Fig. 63 shows its star-shaped
section in actual size. It is crushed to
less than half its thickness by the pressure of the two half-inch bolts screwed
up easily by a man of ordinary strength
with twelve or fourteen inch wrenches.
The bolts are left-and-right threaded. Two ratchet-
wrenches working in opposite directions are used
to correspond with the reversed threading of the
bolts. The pressure exerted by one wrench is thus
resisted by the other. This avoids the necessity of
securing the pipes while the nuts are being screwed
up, and causes both sides to be compressed alike,

since the wrench which has given and received the greatest pressure ceases temporarily to turn until the other has caught up with it.

The " Sanitas " wrenches made for this pipe are so formed that the joint may be made in the most

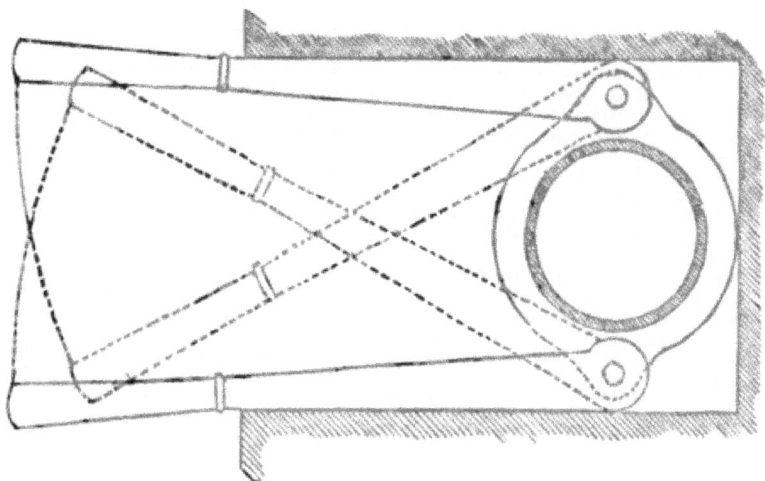

Fig. 64·

contracted places, as shown in Fig. 64, where the stack stands at the bottom of a slot a foot deep and only eight inches or one brick in width.

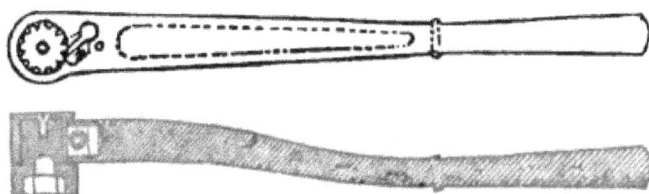

Fig. 65. "Sanitas" ratchet-wrenches.

Fig. 65 shows the ratchet-wrenches.

By the use of these ratchet-wrenches the joint may be thoroughly calked by a single ordinary unskilled workman, after the pipes are once set in place, *in less than twenty seconds*. To calk an ordinary bell-and-spigot joint in the usual defective manner is estimated by good authorities as requiring on the average, when the pipes are once in place, *as many minutes.* The amount of lead used for calking our flanged joint is about one eighth that required for the ordinary joint. The lead gasket for four-inch pipes weighs half a pound ; and for two-inch pipes, one fourth of a pound, while the rule for calking ordinary joints is to use one pound of lead for every inch in the diameter of the pipe. We also save the fuel, oakum, etc., used in making ordinary joints, and avoid the danger of lead-melting in houses.

Figs. 66 and 67 show in section and elevation the simple method of connecting lead and iron pipes when the flanged joint is used. With ordinary bell-and-spigot pipes the proper connection between lead and iron is both laborious and expensive, re-quiring the use of brass ferrules. The lead pipe has to be wiped on to the brass ferrule, and the brass must be calked into the iron hub. A double joint is thus required, and this, especially with the larger pipes, in-volves the use of considerable skill

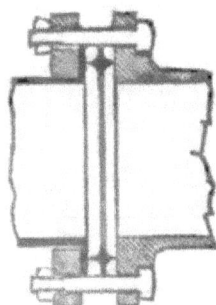

Fig. 66. Section.

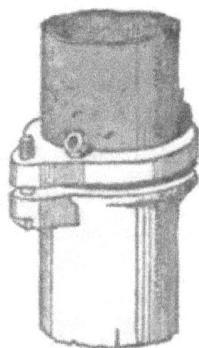

Fig. 67. Perspec-tive view.

and valuable material. With our flanged joint all this is done away with. The lead is simply flanged out to correspond with the flanges of the iron pipe to which it is to be connected, and bolted to the pipe by means of a cast-iron ring furnished with the pipe-fittings, and having ears and bolt-holes corresponding with those of the pipe-flanges. The lead packing-ring is used between the lead and iron flanges exactly as if the flanges were both of iron. In this manner a permanent steam-tight joint is formed between the two metals without hand-calking, brass ferrule, or joint-wiping.

In bell-and-spigot pipes comparatively few bends and branches are made. Should the angle required to reach a certain fixture in laying the pipe be a different one from that given by the bends furnished, the desired direction must be obtained by canting the spigot slightly in the socket, a movement different from the axial rotation we have already described and provided for. To accomplish the same result with flanged pipes a greater variety of castings are made, furnishing bends of a larger number of angles, in the same manner as is done in wrought-iron piping when used with screw-joints for plumbing purposes. With the flange-joint a certain play is obtained by screwing up that side of the pipe upon which the greatest inclination is to be given slightly more than the other. But the variety of castings furnished enables every requirement to be met, without resorting to the method of unequally compressing the packing-ring. We find $\frac{1}{4}$, $\frac{1}{6}$, $\frac{1}{8}$, $\frac{1}{16}$, $\frac{1}{32}$, and $\frac{1}{64}$ bends corresponding to angles of 90°, 60°, 45°, 22$\frac{1}{2}$°, 11$\frac{1}{4}$°, and 5$\frac{5}{8}$° respectively. By using some one of these bends, or a combination of two or more, any desired direction can be obtained. The half-rings are required only for bends and branches. Straight pipes are screwed together directly, and have ears and bolt-holes at both ends. The straight pipes are manufactured in lengths of 1ft., 2ft., 3ft., 5ft., and upward; 10 inches, 9 inches, 8 inches,

7 inches, and 6 inches. Pieces having a bell or a spigot on one end and a flange on the other are also made
for connection with old bell-and-spigot pipes, or for
substituting bell-and-spigot pipes at any point desired in case suitable " Sanitas " fittings cannot be
obtained at momentary notice or in out-of-the-way
places. Moreover, by using these connections, any
one desiring to try the " Sanitas " pipe for the purpose of comparing it with the ordinary pipe before
using it exclusively can incorporate in his piping a
few lengths of the " Sanitas " pipe, and observe its
action under hydraulic pressure and otherwise, as
compared with that of the ordinary pipe, of which
the remainder of his work may be composed. Although this joint requires less space for setting than
any other, it is still always best to give ample room
for it, especially if it be set in slots, and particularly if the plumber is accustomed to the old bell-
and-spigot pipe only. It is recommended never to
set soil-pipes in slots. But if it must be done, the
slots should not be over four inches deep or less than
one foot wide.

As already explained, the hydraulic test, which
should in every house be required before the work

Fig. 68. Capped
end.

can be pronounced safe, is on ordinary bell-and-spigot pipes very difficult of application, because there is
no easy method of temporarily closing
the outlets. Here again our flanged
joint presents an advantage of great
importance. In order to close the
opening it is only necessary to screw
on caps provided with ears and bolt-
holes corresponding with those of the pipe-flanges,
as shown in Fig. 68. The regular packing-ring is
used between the cap and flange, so that the joint

is steam-tight, like the rest of the piping. When the test has been made the caps can be removed and used again and again by the plumber. They are furnished with the rest of the pipe and fittings. The lead rings after use can be used for old lead or recast into new rings.

It is sometimes required in practice that each pipe used be tested at the foundry before coating it, in order to ensure soundness. With ordinary bell-and-spigot pipes, the application of a pressure test is difficult, if not impossible. The straight lengths can be tested under pressure, but the branches and bends offer serious difficulties on account of their form. Hence the oil test has to be resorted to, and the strength or thickness of the pipe is not by this method made known; moreover, the oil test is in many other respects obviously inferior to a strong pressure test. A simple machine has been devised to test these flange-pipes. It consists of two plates with rubber disks on one side, which are pressed against the flanges of the pipe to be tested, by means of clamps and wedges of peculiar form, designed for rapid application. One of the plates is perforated and connected with a water-pipe and pressure-gauge. A simple force-pump is added, so that where the water-pressure is subject to considerable fluctuation, each pipe may be tested under precisely the same pressure. By the use of this device, flanged pipes of any desired size, and all the branches and fittings, may be quickly and accurately tested at the foundry before coating.

In the case of ordinary bell-and-spigot pipes, the expense and imperfections of the jointing are so great that the pipes are cast very long, in order to save joints as far as possible. The attempt to cast pipes of small diameter, say two-inch, three-inch, and four-inch, in lengths of five feet, is almost certain, unless special precautions are taken, to result in an inequality in the thickness of the metal. The writer has found bell-and-spigot pipes of five-feet lengths, made by the best firms and sold for extra heavy weight, no thicker than a piece of thick paper on one side and half an inch on the other. Fig. 69 is an accurate drawing of a two-inch pipe which he has recently been obliged to reject, among a large number of others from the best makers, upon testing them before they were laid by the plumber. It is much more unusual to find pipes of equal than of unequal thickness throughout. This is a very important consideration. The strength and thickness of a line of piping is equal to its thinnest part, as the strength of a rope is equal to its weakest part. Hence all the metal used in the piping of a house beyond the thickness of its thinnest part is thrown away. Of what use is it to pay for extra heavy pipes, when one side of most of them is extra light? It is not for the strength of the piping that we require the thickness, since they are not used like columns to support floors and walls, but for security against leakage and decay. Now since, as is very well known by plumbers and engineers, the majority of long pipes of small sizes are uneven in thickness, the chances of obtaining only the even pipes throughout an entire stack are obviously infinitesimally small, and it is not proba- ble that one house in a thousand exists in which one or more of the lengths of pipe are not very seriously uneven. The enormous waste of metal and the great danger of leakage which this condition of

Fig. 69. Imperfec- tions of long castings.

things implies renders it of the utmost importance to **employ some** means of remedying this great defect.

We find **a** remedy in using **short** castings as **far** as possible, and in casting the long pieces with unusual precautions and in **a** different manner from **the** short ones. The **plumber would** find **great** advantages to offset the **inconveniences in using a** variety **of short** lengths **of pipes,** instead **of** frequently cutting the usual **five-feet** lengths **to fit the** spaces **between** the floors **and** fixtures. Cutting **cast-iron is an** extremely difficult and tedious process. **Were** these various **lengths** manufactured **from six** inches upward, he **would find it** possible to avoid cutting entirely, **and probably** add very few joints **to** the **number now required, for it must be** borne **in mind that** each **time a** pipe is cut **a new** joint is necessitated, **so that** the saving **in the number** of joints **in** ordinary plumbing practice, **by** using **no** other **than** five-feet castings, is much **smaller** than is at first supposed. Now, however, **that we** have found **a** simple, safe, and economical **joint to take** the **place of** the clumsy, uncertain, and expensive **one in vogue, we** have **no** further need of long castings. **The saving in** pipe-cutting, to say nothing of the other advantages, far more than offsets the labor **of** making an extra joint **or** two, and we have **a** stack of pipes whose thickness can be relied **upon** as being uniform throughout.

Where **it is found** necessary for any purpose to remove a **piece of pipe** from a stack already set up **it is** only **necessary to place** temporary supports under the pipe **above the one to be** disjointed, unscrew the bolts, **remove one of the lead** rings by means of a chisel or saw, and **slip out** the length to be removed.

To replace a pipe or fitting several methods may be employed, of which the best is that in which short flanged or threaded brass pipes are used. To the lower end of the pipe to which the new piece is to be connected is bolted a piece of short flanged brass pipe. Another short flanged brass pipe is then slipped over the first, being made just large enough to do so, and the fitting to be connected is afterward bolted to its flange, and to the main piping below. The two brass pipes are finally connected by means of an ordinary wiped solder joint. Or brass pipes may be screwed to iron flanges, to save the more expensive metal.

Another method is to substitute iron for the brass pieces, one of the pices being provided with a hub, and calk the joints by hand in the usual manner. This latter method, however, has the objections of all hand-calked joints, and is for this reason not to be recommended when the first can be applied. By this means the flanged pipes may be connected with old work in which bell-and-spigot pipes have been used. Or bell-and-spigot pipes may at any place be inserted in the line of flanged pipes in this manner, if desired.

As we have already explained, the bell-and-spigot joint is incapable of withstanding the effects of sudden and severe variations of temperature. The spigot being nearest the heat expands more than the hub and compresses the surrounding lead, permanently diminishing its bulk, and forming a passage for the escape of gas. The principle of the construction of our flange-joint is such that this trouble is overcome. The flanges are affected equally by changes of temperature, and the lead packing is never compressed by expansion or contraction. Thus supposing, when the pipes are cool, steam is suddenly allowed to pass through them. Both upper and lower flanges and the lead firmly imbedded between them expand alike outward under the same degree of heat, and return again unaltered as the pipe recools.

The bolts expand and contract with the changes of temperature proportionally with the flanges, and do not affect the packing.

To give these theories a practical test, I have had some four-inch piping connected and closed up at the ends with our flanged joints, and coupled the whole with the boiler of a steam-engine, the steam-gauge indicating about thirty pounds pressure. The steam was left on until the pipe-flanges and bolts had all become thoroughly heated through. The coupling was then immediately transferred to the cold-water supply from the city main, and after the steam had been let out the cold water was suddenly turned on until the piping was filled. As the experiment was performed in midwinter, the test was as severe as possible. The cold water was then poured out and steam again immediately applied. This alternating application of steam and cold water was repeated successively a dozen times. During the entire process no sign of a leak either of steam or water was obtained. The bolts had been screwed up in the ordinary manner without extra care.

It is well known that no bell-and-spigot joint will stand such a test even after the most careful calking.

The same variations of temperature cause the pipes to expand and contract also longitudinally. But in this direction there is always ample play left in setting the pipes for this action, and the lead is obviously not affected by it. Each packing has

upon it the weight of all the pipes above it, as well
as the pressure exerted by the bolts. The weight is
therefore constant, whatever be the temperature or
length of the pipes, provided they are properly set.
The expansive force of iron is so great that if free
play is not allowed for it in a building, it will make
way for itself by tearing away its bonds. Mr.
Bayles says: " In setting up a line of soil-pipe,
intelligent provision should always be made for expansion and contraction of the metal resulting from
changes of temperature. These changes, however,
are seldom sudden or extreme; but when the pipe
is at any point rigidly fastened to the wall it expands in both directions. The amount of motion
at the ends is small, but it must be provided for, or
it will provide for itself. The power with which
iron expands, as its temperature is raised, is practically irresistible. The end of a pipe may not
move more than an eighth or sixteenth of an inch,
but the power with which it moves that distance is
so great that it can only be resisted by a power
great enough to crush the metal. This would be, in
ordinary cases, equal to about 75,000 pounds per
square inch, the strength of cast-iron to resist
crushing strains being from 60,000 to 90,000 pounds
per square inch. Consequently, we see that unless
the fastenings at the ends of a line of cast-iron pipe are
of such a character as to admit of slight movement,
something must give way, and it is not likely to be
the pipe. This, then, must be provided for in the
character and position of the fastenings, which
must be so arranged that, while allowing for some
movement, they shall not develop a tendency to
break or loosen the joints. Under ordinary conditions the amount of expansion is seldom great
enough to give much trouble, but when steam or a

great volume of very hot water washes into an iron pipe it is sometimes great enough to loosen joints and even crack the pipe."

Accordingly, if a line of pipe is rigidly fixed at the bottom, the hooks which hold it against the walls should be placed a short distance away from the flanges, so that the line of piping is free to slip up and down slightly under the influence of expansion and contraction. Otherwise these hooks are liable to be loosened from the mortar or woodwork into which they may be driven, since it would be easier for the pipe to loosen the hooks in the mortar or wood than to further compress the packing-rings, or to stretch out the heavy bolts of wrought iron.

The opportunities furnished by the ordinary bell-and-spigot joint for careless or fraudulent work are avoided in our flanged joints. The entire thickness of the lead is visible from the outside between the flanges. As the lead is the only packing used, and as this is in open view, nothing can be fraudulently omitted. The bolts and nuts are also visible and, moreover, must be of the standard size and strength in order to furnish the requisite amount of compression to stand the hydraulic test.

Thus we find in our " Sanitas " joint all the characteristics demanded for plumbing purposes.

1.   It is water, gas, and steam tight, even under heavy pressure.

2.   It is unaffected by the expansion and contraction of the pipes.

3.   It is capable of resisting severe jars and strains, both compressive and tensile, such as are occasioned by the weight of the pipe, or by settlement and movement in the building.

4.   It requires neither skilled labor nor machinery in its manufacture or in its jointing.

5. It is of such a form and nature as to admit its being as easily taken apart for repairs or alterations as it is put together, and this without damage to any part.

6. Its form and construction are such as to allow it to be made and put together rapidly, to follow easily the irregular contour of the construction, and to be used immediately after fixing in place.

7. It requires no hand-calking or hammering, which are liable to fracture the pipe or its lining.

8. It is so formed that any imperfection, either in the materials used or in the manner of putting them together, can easily be detected at a glance from without, without expert aid.

9. It is compact enough to permit its use in the most contracted spaces.

10. It causes no obstruction to the water-way, and leaves no appreciable space or pocket for deposit.

11. It is simple, durable, indestructible, economical, and unobjectionable in appearance.

It is, therefore, suitable for water, gas, and steam under pressure, as well as for drain and soil pipes.

SIZE AND GENERAL ARRANGEMENT OF THE PIPING.

HAVING thus described a safe and economical method of jointing our cast-iron pipes, it remains to consider their proportions and general arrangement. We shall, as treating of house drainage, confine our attention to the piping of the house proper; the consideration of the drainage beyond the house limits belonging more properly to the subject of sanitary engineering.

The size of soil and drain pipes should not exceed four inches. This is ample to carry off every possible form of discharge or combination of discharges to be met with in plumbing, even in the largest buildings, except for special hotel, laundry, or manufacturing purposes. It is a mistake to suppose that because the fixtures are multiplied, the diameter of the soil-pipe must be multiplied correspondingly. It is a rare occurrence, even in a hotel-building, that several water-closets are flushed at exactly the same instant, and even if they were, their distance from each other and the capacity of a four-inch pipe would give ample room for the escape of the water. The choking of a pipe is far oftener due to its being too large, or to faulty construction, than to its want of sufficient size. The great perpendicular extension of plumbing-pipes also facilitates the discharge, and pipes which for land-drainage would be much too small, will be found ample for plumbing purposes on this account.

When we consider how important it is that the soil and drain pipes should be as thoroughly scoured as possible by the discharges sent through them, and remember that the smaller the pipe the more perfect the flushing, we should be inclined to reduce our soil-pipes to a size even smaller than four inches, were it not for the careless usage slop-hoppers and water-closets are so often subjected to.

The traps of water-closets are very frequently made of pipe as much as four inches in diameter, and large pieces of newspaper are often used where toilet-paper alone is suitable. This, and the carelessness of servants who will throw into a closet anything which is small enough to pass through its trap, would cause a great amount of annoyance and expense if the soil-pipe were smaller than the trap of the slop-hopper or water-closet. Hence we have fixed upon four inches as both the smallest and the largest size of soil and house-drain pipes, and believe that no other size should be used except for rare and exceptional cases.

All the piping of a house should be in full view. Nothing should be walled in or covered over and rendered inaccessible. One of the first rules of modern sanitary work is to bring everything out of the darkness into light and air, where defects, if they occur, can at once be detected and removed. We are accustomed to running our steam-pipes in plain sight, and rendering them, by gilding or silvering, as ornamental as possible. The same custom is now beginning to apply to our plumbing-pipes. Where they pass through parlors or reception-rooms, they should stand behind movable panels or doors : a little ingenuity on the part of the architect will generally enable this to be done with good effect.

The piping should be arranged to run as direct as possible, and should be concentrated. It will be found very convenient, especially in city houses, to build a broad recess, or slot, in the masonry of the party-wall, on the line of the bath or toilet rooms, for all the plumbing and ventilating pipes which can be collected together in this neighborhood, and, if possible, to run up in this slot the smoke-flue of the furnace, in iron. The heat of the smoke-flue will create a strong circulation in the ventilation-pipe, and at the same time radiate a useful heat into the bathrooms. The brick recess should be enclosed in masonry on all sides where it passes through the floor, and as high as three or four feet from the ground. This serves to protect the woodwork from danger of overheating. Between this height and the ceiling the recess is open in front, exposing the pipes and admitting the radiant heat from the flue. The various stories are separated from each other by brick platforms, built across the recess on the line of the floors, and made tight around the pipes with cement or mortar. Above the upper bathroom the iron smoke-flue enters a regular brick flue, and the soil-pipe ventilator runs up independently through the roof. The writer has adopted this system in several city houses and found it very satisfactory. The furnace smoke-flue may be constructed of tile instead of iron, if preferred, for greater durability ; but if iron be used, it may be made heavy enough to last as long as desired. The recess being, moreover, accessible, the pipe may be renewed at any time without difficulty.

Every stack of soil-pipe should be thoroughly ventilated, by being extended full-size from the bottom to, and through, the roof. No ventilating-pipe running through the roof should be of less

diameter than four inches, inasmuch as smaller
pipes are liable to become clogged in winter by
snow and frost.

The extensions above the roof should not be less·
than two feet high and the tops should never open
near a chimney - top, ventilating - shaft, dormer-
window, or other opening, for obvious reasons. It
is generally sufficient to allow the pipes to remain
wide open, without return-bends or ventilating-caps,
which only serve to obstruct the circulation. Wire
nettings may be put over the opening at the top, to
prevent objects from falling into the pipe, through
accident or malice.

These iron ventilating and soil pipes form the
best possible lightning-conductors, because they
are always sure to have a good and moist ground-
connection, and are composed of a body of metal
heavy enough to carry the most powerful charges
of electricity without danger of melting ; their
presence, therefore, in sufficient number, renders
the usual form of lightning-rod superfluous.

The soil-pipe should be firmly suported at the
bottom. The best support consists in the projection
of the foundation-wall, or in a stone or brick pier
made for that purpose. The junction between the
soil and drain pipes should be made with an easy
bend, of as large radius as possible, to prevent the
accumulation of obstructions and the powerful back-
pressure on traps caused by the friction of the air
in attempting to pass round a sharp bend in front of
a descending column of water in the soil-pipe.

The main-drain should run along in full view on
the foundation-wall, if possible, or supported by
piers resting on the concrete, or hung from the
joists by strong iron hangers. Clean-out openings
should be provided at all places where sediment or
obstructions are liable to collect.

Sometimes it is found convenient to rest the drain directly on the concrete. In this case it is customary to form the concrete in a trench whose bottom pitches with the proper grade to accommodate the drain. The drain should have a fall of half an inch to a foot, if possible, or at least a quarter of an inch to a foot; of course, the greater the pitch the better.

The main-drain should be trapped with a running-trap of iron just inside the cellar-wall, or, if this is impossible, outside the house, in a manhole. The trap should always be accessible and should be provided with clean-out caps with air-tight covers. It is a good plan to run a water-conductor into this trap, to ensure its occasional flushing.

To provide for a complete circulation of air through the soil and drain pipes, a fresh-air inlet of the full size of the drain should enter it just inside (on the house side) of the main-trap above described. The mouth of this inlet should open outside of the house, at some little distance from any door or window.

Where a fixture is connected with the rigid iron soil-pipe stack, provision must be made for a certain degree of movement or play on the part of the fixture, in such a manner that the movement shall not crack the joint or in any way loosen it. A settlement of the masonry, a jarring or shrinkage of the floors is certain in a new building to alter to a greater or less extent the relative positions of the fixture and its soil-pipe connection. Injury to the joint from this cause may be prevented in two ways. One of these is to use a sufficient length of lead pipe in all cases between the fixture and the iron stack, and the other is to support the fixture directly on the stack itself, and make it entirely independent of the floors and woodwork.

The first method is used with cast-iron piping and with all fixtures having waste-pipes of small calibre. Lead piping has so much flexibility that a section of even moderate length will permit of a considerable movement on the part of the fixture without injury to the joint. Where the fixture is a water-closet, a length of lead pipe having a horizontal extension of two feet, or its equivalent, in any inclined direction, will permit the utmost shrinkage of joists or settlement of walls liable to occur in good plumbing after the plumbing is connected, without injury to the work.

The second method, that of supporting the fixture directly on the stack, has been successfully employed with wrought-iron piping. The closet sets on a cast-iron base firmly attached to the piping, so that a shrinkage of the floor may take place without affecting the joint.

It is important that all angles and bends in our piping should be as smooth and gradual as possible.

No sharp angles should be allowed. Thus, if it is ever necessary at any point to use T-connections, they should not be formed as they are in common bell-and-spigot fittings, but should be made with a curve at the branch junction. But Y-branches are to be preferred to T-branches, and it is very seldom, if ever, that T's are required.

The peppermint and smoke tests are useful for application at any time after the house has been occupied, when it is desired to ascertain if the pipe system has remained sound throughout, especially in places where a leakage of water might not occur. To apply the peppermint test, the vent-openings are first to be closed with plugs. A two-ounce bottle of oil-of-pepermint is then carried up to the roof by an assistant, and its contents poured into the soil-pipe at its mouth above the roof. A pail or pitcher of hot water is immediately poured down after it, and the opening is then plugged up. The assistant remains upon the roof until the examination within the house has been completed ; otherwise the odor clinging to his clothes will be likely to follow him into the house. The peppermint is volatilized by

Fig. 70. Asphyxiator for applying smoke test.

the hot water, and should any leak occur it will at once be detected and located by its pungent odor, unless the pipes have been improperly imbedded in the walls or are so covered up that access to them is impossible. If they are set as they should be, everywhere in open view, no difficulty will attend the detection and repair of the minutest defect or leak. The smoke test is applied by means of special bellows manufactured for the purpose (Fig. 70). It enables those whose sense of smell is not acute to operate instead with the sense of sight.